✱ DEFENDERS OF THE RACE ✱

DEFENDERS OF THE RACE

JEWISH DOCTORS AND RACE SCIENCE
✣ IN FIN-DE-SIÈCLE EUROPE ✣

JOHN M. EFRON

YALE UNIVERSITY PRESS
NEW HAVEN AND LONDON

Copyright © 1994 by Yale University.
All rights reserved. This book may not be reproduced, in whole or in part, including illustrations, in any form (beyond that copying permitted by Sections 107 and 108 of the U.S. Copyright Law and except by reviewers for the public press), without written permission from the publishers.

Designed by Nancy Ovedovitz and set in Sabon type by Rainsford Type.
Printed in the United States of America by BookCrafters, Chelsea, Michigan.

Library of Congress Cataloging-in-Publication Data
Efron, John M.
Defenders of the race : Jewish doctors and race science in fin-de-siècle Europe / John M. Efron.
p. cm.
Includes bibliographical references and index.
ISBN 0-300-05440-8
1. Jews—Identity. 2. Race. 3. Physical anthropology—History. 4. Jewish scientists. 5. Antisemitism. I. Title.
GN547.E34 1994
572—dc20 94-10413
CIP

A catalogue record for this book is available from the British Library.

The paper in this book meets the guidelines for permanence and durability of the Committee on Production Guidelines for Book Longevity of the Council on Library Resources.

10   9   8   7   6   5   4   3   2   1

FOR MY PARENTS

שמע בני מוסר אביך ואל תטש תורת אמך

כי לוית חן הם לראשך וענקים לגרגרתיך

✣ CONTENTS ✣

List of Illustrations
ix

Acknowledgments
xi

One
Introduction
1

Two
German Race Science: The Jew as
Essential Other
13

Three
British Race Science: The Jew as
Non-Other
33

Four
Joseph Jacobs and the Birth of
Jewish Race Science
58

CONTENTS

Five
Samuel Weissenberg: Jews, Race,
and Culture
91

Six
Zionism and Racial Anthropology
123

Seven
The Limits of Racial Self-Representation
175

Notes
181

References
215

Index
243

ILLUSTRATIONS

(following page 90)

"Jewish Face." From Maurice Fishberg, *The Jews: A Study of Race and Environment* (New York, 1911).

"The Jewish Type: Composites." From Joseph Jacobs, *Studies in Jewish Statistics* (London, 1891).

"Young German Jews with Negroid Features" and "Young German Jews with Nordic Features." From Sigmund Feist, *Stammeskunde der Juden* (Leipzig, 1925).

"Types from Baluchistan with Jewish Physiognomy" and "Japanese with Jewish Physiognomy." From I. M. Judt, *Die Juden als Rasse* (Berlin, 1903), and Maurice Fishberg, *The Jews: A Study of Race and Environment* (New York, 1911).

"South Palestinian Jewish Type." From I. M. Judt, *Die Juden als Rasse* (Berlin, 1903).

"Jews from Kai-feng-fu [1902]." From Sigmund Feist, *Stammeskunde der Juden* (Leipzig, 1925).

"Theodor Herzl with Matvei Bogatyrev and Salomon Marduchaev, representatives of the Caucasian Mountain

ILLUSTRATIONS

Jews at the Sixth Zionist Congress, in Basel, 1904." From Sigmund Feist, *Stammeskunde der Juden* (Leipzig, 1925).

"Jews, Teutonic Type, United States" and "Polish Jew in Jerusalem." From Maurice Fishberg, *The Jews: A Study of Race and Environment* (New York, 1911).

"Jewish Girl from Tunis." From C. H. Stratz, *Was sind Juden?* (Vienna, 1903).

❊ ACKNOWLEDGMENTS ❊

Over the long period of time it takes to complete a book one accumulates significant debts of gratitude. While these can never be fully repaid, a step towards partial restitution may be made at this point.

First and foremost comes my family. In creating a home rich in Jewish culture and feeling, my parents, Miriam and David Efron, provided me with a sense of the relevance of tradition and the past. Their unflagging dedication to my chosen career, and the personal sacrifices which that entailed on their part, are expressive of their love. It was they who provided me with the incentive and the means to embark upon this protracted journey so far from my native Australia.

If it was my parents who encouraged me to begin graduate work in history, it is thanks to my beloved wife Deborah that I finally became an historian. Her cheerfulness and understanding, as well as her patience and forbearance, have provided me with sustenance ever since we first met on a warm Tel Aviv evening in 1980. To repay her with the same dedication and loyalty while she pursues her own career, so selflessly put on hold for the sake of mine, is my goal. The writing of this book was made all the more pleasurable by the welcome distractions offered by our beautiful children, Hannah Elka and Noah Oliver.

A long line of teachers and colleagues have left their imprint on both me and this work. Professors Yosef Hayim Yerushalmi and Nancy Stepan, at Columbia University, taught me the historian's craft. Professor Yerushalmi's profound knowledge and mastery of the long history of the Jewish people are a source of inspiration. His intellectual rigor, demanding standards, and compassion make him the exceptional and exemplary teacher he is. I thank him deeply for his support and for his faith in me. To Professor Stepan I shall always be grateful for having introduced me to the history of science. From our very first meeting she took an abiding interest in my work, and her steadfast encouragement never once waned. Her infectious enthusiasm for both scholarship and teaching, coupled with her exacting standards, have left an indelible impression on me.

I also wish to thank Sander L. Gilman, who provided wise counsel every step of the way. His path-breaking work on medical images of the Jews lay the basis for my own study. I also wish to express my gratitude to George L. Mosse, my teacher at the Hebrew University of Jerusalem, whose intellectual influence upon me is stamped clearly on this book. Yehuda Bauer, also at the Hebrew University's Institute for Contemporary Jewry, has played an important role in my development as both a scholar and teacher. Michael Meyer also generously gave of his time and expertise, reading the entire manuscript and offering valuable criticisms and encouragement.

It is a delight to be able to acknowledge two dear friends who have made important contributions to this study. My friendship with David Myers is a special joy and privilege. As a source of constant intellectual challenge and pleasure, as well as in innumerable other ways, he has made a welcome impact on me professionally and personally. My thanks are also due to Michael Brenner, who with unfailing grace has been ever prepared and willing to share with me his knowledge and expertise. In our ongoing conversation, I continue to learn much about the Jewish experience in Germany, both past and present.

I also wish to express my appreciation to Charles Grench of Yale University Press for his support of this project and to Richard Miller, the manuscript editor, for his sense of style and intellectual engagement.

My graduate work, as well as research trips abroad, was made possible through a number of grants and fellowships from several sources. I received financial support from the Memorial Foundation for Jewish Culture, the Lady Davis Fellowship Trust, the German Academic Exchange, and the Center for Israel and Jewish Studies at Columbia University.

✣ DEFENDERS OF THE RACE ✣

ONE

INTRODUCTION

> Es ist das beharrlichste Volk der
> Erde, es ist, es war, es wird sein.
> *Goethe*, Wilhelm Meisters Wanderjahre

A large part of contemporary intellectual, political, and indeed popular discourse revolves around issues of race and ethnicity. Despite the seeming newness of these categories and modes of self-perception, they first made their appearance in the eighteenth century under the aegis of Romanticism and gained additional importance during the late nineteenth century under conditions of intense nationalism, imperialism, and xenophobia.

In contemporary debates on race, minority voices are at last being heard. Minorities help guide the discourse and are not merely presented as abstractions or anthropological curiosities as once was the case. Inspired by the profound changes stemming from the social and political unrest of the 1960s and the subsequent conservative backlash of the 1980s, the work of scholars concentrating on the history of oppositional strategies of minority groups has tended to focus on social and political resistance to marginalization and discrimination. Rarely have scholars sought to document modes of intellectual resistance to prejudice. Specifically, students of such forces have been especially slow to trace the resistance strategies of minorities within the various branches of science and medicine.[1]

The reason for the absence of such scholarship lies in the image of science as something objective and value-free. Faith in its universal language and application has led scientists themselves to dismiss or even sweep under the rug episodes in the history of science that constitute a threat to science's "good" reputation.

One such threat comes from scientific racism. In part because of Nazi atrocities in the name of "race," scientists and historians have tended to view race science as a pseudo-science. By definition, therefore, studying a pseudo-science has been regarded as an intellectual indulgence, despite the central place that notions of race have occupied in modern culture, and not only that of the West. To study the few groups who were able to mount a sustained response to the scientific discourse on human variation has been regarded as an even greater luxury and has therefore not received the attention such inquiry deserves.[2]

Self-assertion, and this is what the contemporary discourse on ethnicity is about, is the result of and made possible by greater power sharing and, especially, increased access to knowledge.[3] However, the novelty of this situation vis-à-vis minorities is not entirely without historical precedent. Cultural anthropologists have recently been drawing attention to the loss of ethnographic authority in social anthropology. These critics have noted that no longer do Western participant-observers possess an unchallenged voice when it comes to describing the Other.[4] If it is true that cultural representation is contestable, historical, and constitutes a rhetorical performance, it is my contention that the same holds true for the narratives of physical representation. But when analyzing the texts of the Jewish physical anthropologist in particular, one is forced immediately to confront and challenge the periodization established by the cultural anthropologist. James Clifford argues that the "breakup of ethnographic authority . . . is linked to the breakup and redistribution of colonial power in the decades after 1950 and to the echoes of that process in the radical cultural theories of the 1960s and 1970s." He further asserts that "after the negritude movement's reversal of the European gaze, after anthropology's crise de conscience with respect to its liberal status within the imperial order, and now that the West can no longer present itself as the unique purveyor of anthropological knowledge about others, it has become necessary to imagine a world of generalized ethnography."[5]

By locating the origins of a challenge to the dominant ethnographic discourse only in the period after World War II, cultural critics of anthropology have been led astray, conceiving of the enterprise of anthropology merely as one in which white Europeans have gone into the field to observe other, distinctly different races. What modern critics have overlooked is that in the nineteenth century, anthropology classified the Jews as a race. As this book will demonstrate, the Jews were not easily pigeonholed, and their racial classification could differ from one anthropologist to the next. But the basic assumption, that the Jews were radically different from their Christian neighbors, led Europeans to see them as a race apart. Once we recognize that nineteenth-century anthropology identified literally scores of nationalities and ethnic groups within nation-states as distinct races, and that the observed could even resemble the observer, it becomes possible to revise Clifford's periodization when it is acknowledged that nineteenth-century Jews, as representatives of a distinct racial group, mounted a challenge to the dominant ethnographic and anthropological canon.

The existence of Jewish race science texts from the fin de siècle calls into question the notion that postwar European decolonization was the first challenge to the hegemonic discourse of European anthropology. Moreover, to allow fully for this notion, even the categories of "empire" and "colonized" need to be expanded to include groups such as Jews, who do not fit neatly into the traditional paradigm of empire, as that term is understood by anthropologists and historians. Doing so, however, would make possible a revision of when the breakdown of European ethnographic authority began. Anthropologists and historians of anthropology have viewed fieldwork as the action of a subject (European scientist) upon an object (exotic aborigine), by which the subject reaches out beyond the confines of a metropolitan center toward a peripheral outpost and at the same time reinforces the boundaries between the two.

To accept the proposition that historically the Jews were involved in a colonial relationship with Christian Europe is to also recognize that the labors of Jewish physical anthropologists were an attempt at reversing the European gaze. In late-nineteenth- and early-twentieth-century Europe, access to knowledge allowed Jewish scientists to engage the dominant discourse about race and the so-called Jewish question as well as to mount a sustained campaign of self-defense, self-assertion, and ethnic identity building.

At the end of the nineteenth century, the terms "Jewish racial question," "Jewish question," and "Jewish problem" were used interchangeably by Jews and non-Jews alike. All these terms refer to aspects of Jewish society, or of Jews themselves, that were considered to be objectionable and in need of improvement. They were broadly defined to include the religion, culture, economic status, and political inclinations of European

Jewry. In this book I approach the phrase "Jewish racial question" literally, as part of the scientific discourse that was concerned with the biological properties of the Jewish people. However, as will become apparent, fin-de-siècle notions of race impinged upon questions of history, so that even a purely scientific approach to the Jewish racial question meant the necessary introduction of discussions about the other categories of the Jewish problem.

An imperative of the Jewish experience in the modern era has been the need to repeatedly redefine Jewish identity in the wake of emancipation. It is not the place here to enter into a detailed discussion of the modes of modern Jewish self-definition, other than to say that they varied in accordance with an ever-changing intellectual, political, and social zeitgeist. In Germany, for example, Jews went from being called and calling themselves Jews in the eighteenth century, to Israelites in the nineteenth, to German Citizens of the Mosaic Faith into the twentieth.[6] Such nomenclatural change among minority groups is not unusual. In the United States, for example, since World War II, one segment of the population has at various times been referred to by itself and others as Negroes, coloreds, blacks, and, of late, African-Americans. To return to the Jews (but with obvious modifications the following could also apply to the above American example), all the names that they employed to refer to themselves reflected changing ideas of Jewish self-perception and varying notions of the Jews' relation to the host nation.

The mode of Jewish self-representation that I focus on in this book is just one of many in widespread use from the late nineteenth century until the 1930s. It was distinct from earlier patterns in that it moved away from religious definitions, such as Orthodoxy, Positive Historical Judaism, and Reform Judaism, or from mere reductionist ones wherein the national or ethnic dimension of Judaism was denied, allowing those Jews who subscribed to this notion to see themselves as German citizens who were Jews, just as there were German citizens who were Protestants or Catholics. Instead, this mode sought to grapple with an ethnic or racial definition of Jewishness by way of a complex process of absorbing, adopting, and manipulating contemporary scientific discourse about the physical nature of Jewishness.

On one level, then, this book is also about the nature of scientific language. The discourse on race was discriminatory, but nevertheless it was democratically appropriated even by Jews who, in Europe, were the prime targets of racial prejudice. Its adoption permitted Jewish scientists to create a powerful and uniquely modern representation of Jewish identity. To this end, I analyze the work of a number of Jewish physical anthropologists, medical doctors, and ethnographers who between 1882 and 1933 undertook the task of studying the physical and psychological characteristics of world Jewry. My aim has been to study the labors of these Jewish intellectuals who, I believe, attempted to create a new, "scientific" paradigm and agenda of Jewish self-definition and self-perception.

The nineteenth century saw the rise of "respectable" scientific racism and the increasing preoccupation of Jewish scientists with how to respond to the claims of mainstream science that Jews were different from and indeed inferior to non-Jews. From the late eighteenth century to the middle of the twentieth, belief in racial difference was given intellectual prestige and social acceptability by the development of race science, the study of human difference on the basis of supposedly "demonstrable" anthropological, biological, and statistical proofs. As historians of race science have effectively demonstrated, by the nineteenth century the concept of race had become central to the Western intellectual tradition.[7]

Furthermore, race science's very subjectivity made it a versatile tool in the hands of nationalist and colonialist groups, providing scientific justification for aggression against a local people or for lamenting the passing of an age in which groups such as women, Jews, and workers were in their "proper places" and traditional hierarchies were undisturbed. Its development, which coincided with the emancipation of the Jews in central and western Europe, had a profound and specific impact on European Jewry. Race science, often explicitly but perhaps more often implicitly antisemitic, played a significant role in the eventual failure of the emancipation process. The authority of science tended to confirm the common person's opinion. Before scientific racism had run its course, however, Jewish scientists had risen on behalf of their embattled people, polemicizing the problem of Jewish "Otherness" by using the contemporary methodologies of race science to either confirm or disprove claims of racial difference.

Although the period this story covers, the late nineteenth and early twentieth centuries, has been designated by historians as the time of the advent of racial antisemitism, this assessment requires some clarification. Although it cannot be denied that the scientizing of anti-Jewish prejudice took place then, basic premises involving the supposed physical and mental uniqueness of the Jew and the hereditary nature of those characteristics have a much longer history.

In the Middle Ages there was, for instance, a widespread belief that Jewish males menstruated. Its origins were in the work of the thirteenth-century anatomist Thomas de Cantimpré, and it was repeated as one of the accusations against Jews at the Tyrnau ritual murder case in 1494 and by a Jewish convert, Franco da Piacenza, in his book on "Jewish maladies" in 1630. It was held that this pathological uniqueness was a mark of the "Father's curse," a result of the Jews' denial of Christ.[8]

Yosef Yerushalmi has persuasively argued that from the middle of the fifteenth century increasing hostility toward the assimilation and integration of Jewish converts into the highest echelons of Iberian society culminated in the passage of legislation in Spain and Portugal prohibiting *conversos* and their descendants from holding office or receiving privileges and honors. These *estatutos de limpieza de sangre* (statutes of blood purity) are among the earliest known antisemitic race laws.[9]

The claim that the Jews were in some essential way tainted and unique could also be heard from their political friends. In his justly famous *Essay on the Physical, Moral, and Political Regeneration of the Jews* (1788), the pro-emancipationist Abbé Grégoire argued that Jews possessed a number of unique characteristics that set them off from other Europeans. For example, despite his adherence to the Enlightenment concept that the environment has the power to change people, he believed this was not really the case with the Jews: "Climate has scarcely any effect on them, because their manner of life counteracts and weakens its influence. Difference of periods and country has, therefore, often strengthened their character, instead of altering its original traits. In vain has their genius been fettered; it has never changed; and perhaps there is more resemblance between the Jews of Ethiopia and those of England, than between the inhabitants of Picardy and those of Provence."[10]

Citing Shaftesbury's *Characteristics* (1711), Grégoire noted that the Jews appeared "naturally gloomy and melancholy" (53). On the basis of a report from the Swiss physiognomist Johann Caspar Lavater, Grégoire was assured that "the Jews in general had sallow complexions, hooked noses, hollow eyes, prominent chins, and [that] the constrictory muscles of the mouth [were] very apparent" (55). There were sexual charges as well. The Jews were concupiscent because of the "accumulat[ion of] many acrimonious particles in the mass of humours contained in their bodies." Moreover, Jewish women "would be very subject to nymphomania, did they not long pine in a state of celibacy," and Jewish males were chronic masturbators (43).

None of these accusations ever really died out. At the end of the nineteenth century, when antisemitism acquired scientific respectability, these claims were merely metamorphosed into newer, more scientific charges.

INTRODUCTION

No one now believed that Jewish men could menstruate, but it was a commonplace to see the Jews as a whole as an effeminate race.[11] In psychiatry, both Jewish men and Jewish women were labeled as pathologically hysterical.[12] Specific legislation that prevented Jews from attaining a completely equal status in European society was a common feature of Jewish life well after the enactment of the *estatutos de limpieza de sangre* and into the modern period. Grégoire's claim about the libidinousness of Jewish females and the onanistic proclivity of Jewish males not only became a staple of modern antisemitic literature but also underlay medical thinking, particularly in psychiatry, where it was believed that Jews were especially prone to mental illness, partly due to their abnormal sexual practices.[13] Thus, ancient charges were modified and incorporated into both modern medical practice and political administration. The Jewish scientists of the fin de siècle who participated in the contemporary debates over the biological nature of Jewishness were in fact also engaged in answering ancient but persistent antisemitic charges.

Although the work of Jewish scientists writing on race was in part a reaction to the powerful antisemitic forces emerging in both Western and Eastern Europe in the 1870s and 1880s, it was not merely a reflex response. The project of these scientists was propelled by its own peculiarly Jewish inner dynamic. Indeed, their work constituted a novel postemancipatory response to what they recognized as the failure of assimilation and the crisis of Jewish identity.

The antisemitic backlash in European thought and practice that began in the 1870s created a deep sense of shock and dismay among many Jews. Perhaps no group felt more distressed than Europe's acculturated Jewish intellectuals. For those who had pinned their hopes on the nineteenth century as the time when an Enlightenment-inspired, rationalist society would prevail in Europe, the vigorous assault on Jewish life by the antisemites led many to reassess their former allegiance to assimilationism and to abandon any hope of further Jewish integration into European society. Among the many Jews who took this stand were Jewish scientists, primarily in Central Europe.

In Germany and Austria after 1878, the decline of political liberalism and the attendant acceleration of scientific racism and organized political antisemitism spurred Jewish scientists to respond swiftly and vigorously to the various claims of race science and the antisemites. From their studies in the science faculties of the universities and their heavy representation in the medical profession, Jewish scientists were well situated to employ the discourse and methodology of race science and ethnography in order to meet the claims of their opponents. I trace their efforts along two intertwined paths: the work of the principal Jewish physical anthropologists who addressed the charges made against Jews by race scientists, and the broad themes of an anthropology that took as its sine qua non Jewish biological difference.

Jewish anthropologists set out to answer one of the major questions of anthropology at the time: What are the Jews? This problem produced a further, diverse set of questions: Do the Jews constitute a race? If so, do they form a single, stable racial type, or are they made up of many races? Either way, what are their unique characteristics? Are the Jews more susceptible than other racial groups to certain illnesses? Are these dispositions to disease hereditary or environmental? One zoological definition of a race is a group of individuals that have the capacity to interbreed and produce fertile offspring. Scientists, both Jewish and non-Jewish, questioned whether the offspring of Jewish-Gentile marriages were fecund. If they were not, as some claimed, was it because of the pathological difference of the Jews?

Joseph Jacobs, Samuel Weissenberg, Elias Auerbach, Felix Theilhaber, and Ignaz Zollschan all spent the better parts of their professional careers in pursuit of answers to these and allied questions. Their work covered the gamut of concerns in contemporary race science. The staggering output of men like Jacobs, Weissenberg, and Zollschan means that their works form the bulk of all Jewish scientific writings on the Jews as a race. Thus, their studies are also fully representative of the enterprise of all contemporary Jewish physical anthropologists. Although they concerned themselves with a similar set of problems, each approached them from very different cultural contexts, which in turn led to very different kinds of scientific conclusions—an important feature of the scientific enterprise in general. The varying cultural and political environments from which each one came—England, Russia, Germany, and Austria—plus the nature of their own Jewish politics and the positions they occupied on the spectrum of religious belief, played a significant part in the scientific agendas they followed, the way they asked their questions, and the answers they offered. All, however, were engaged in the same enterprise—to redefine Jewish peoplehood in biological and anthropological terms.

These influential men helped define a unique trend within racial science: Jewish resistance. Their labors came at a time in the history of anthropology when that science seemed to have moved furthest away from its Enlightenment origins. Cosmopolitan ideas of universal brotherhood and inherent equality had gone out of vogue, replaced by a biological determinism that saw differences in bodily forms as the key to the unfolding of human history. Jewish scientists could not enter the anthropological debate on Jews and still use the abstract language of the Enlightenment. Instead, they had to appropriate the professional discourse of modern science to challenge the biological and medical evidence brought to bear against them.

This did not mean, however, abandoning their allegiance to those Enlightenment principles that stressed the common descent of humans, their fundamental equality, and their capacity to be effected by environmental forces. It is a crucial feature of the Jewish scientists to be discussed that despite using the contemporary language, and sometimes the methodology of race science, they all thoroughly rejected its use for chauvinstic purposes. This is true not only for Jews committed to the Diaspora, such as Joseph Jacobs, but for Zionists as well.

For Jews, the last two decades of the nineteenth century were dominated by organized antisemitism and massive westward migration. In Germany, the effects of these two historical developments were most keenly felt. From 1878 and the advent of Adolf Stöcker's Christian Social party, organized political antisemitism continued to grow in strength. Public figures joined in the call to exclude Jews from German public life and to ban the admission of East European Jews. By 1893, sixteen candidates had been elected to the Reichstag on a purely antisemitic platform. Antisemites, responding to the success enjoyed by assimilated Jews, agitated vociferously (though unsuccessfully) for the exclusion of Jews from government jobs, the judiciary, and higher education, as well as for an end to all immigration.

The mass migration of some 2.75 million Jews from Russia between 1881 and 1914 led the majority at least to pass through Germany on their way further west. The meeting of German and East European Jews was a most complex and at times painful social encounter which provided grist for the race scientist's mill. The convergence of antisemitism, mass migration, and the East-West encounter put into sharp relief the questions race scientists asked about heredity versus environment. For Jewish race scientists, this situation created an open laboratory whereby race science served as an instrument of self-definition. All the scientists studied in this book were deeply touched by these developments and attempted in their work to satisfy the personal dilemmas they experienced at the unfolding of these events. But they also sought through race science to provide a more universal palliative for the assault on Jewish life in Europe. Their reliance on statistics, measurement, and the quantification of various aspects of Jewish existence was an attempt to find a rational, scientific, and empirical answer to the Jewish question.

In examining the nature of the Jewish racial question in various European countries, I have omitted France. Although it is true that France had a proud scientific tradition and a sophisticated community of physical anthropologists, it contributed relatively little to the discourse on the Jews as a race. The reasons for this have mainly to do with the nature of French antisemitism and the character of French Jewry. Although fin-de-siècle France was home to vicious antisemites such as Edouard Drumont and was the scene of the Dreyfus affair, French antisemitism nevertheless differed from the more pervasive and organized antisemitism in Germany and Austria. In France, antisemitic organizations and parties tended not to exist for the sole aim of combatting Jews but were annexed to broader political causes such as royalism or anti-Freemasonary. In addition, the Jew as the primary Other was displaced after France's defeat in the Franco-Prussian War of 1870 by the Germans. Xenophobia was annexed to revanchism, and the enemy across the Rhine displaced the Jews as the principal target of French hostility.[14]

The Jewish racial question in France was muffled because of the French Jewish community's position in the host society. Whereas Jewish emancipation was a long, drawn-out affair in Germany, with civil rights being granted piecemeal in an atmosphere of ill-will and ambivalence, France's Jews were emancipated in two broad and comprehensive acts soon after the Revolution—first the Sephardic community in 1790, and then the Ashkenazim in 1791. After this time, French Jewry's confidence in the institutions of the French state to ensure the community's security was unshakeable. This attitude of French Jews was most evident during the Dreyfus affair, when despite the frightening atmosphere and anti-Jewish riots, French Jewry steadfastly refused to become actively involved and come to its own defense, claiming that French justice will out. By contrast, in 1893, just a year before the Dreyfus affair erupted, German Jews established a self-defense organization, the Centralverein deutscher Staatsbürger jüdischen Glaubens, in response to the new antisemitism. German Jews simply could not be as confident that the Reich's social and political institutions would protect them as French Jews could be of their country's.

A consequence of the more secure position Jews enjoyed in French society was that they were not a focus of French race science, as the people in the French colonies were. By and large, the Jews of France were regarded as French. Moreover, because of their fortunate position in the French nation-state, French Jews did not develop a tradition of self-

INTRODUCTION

defense as German Jews did. In sum, the Jewish racial question was posed in France only occasionally, and was rarely answered by the Jews of that land.

I begin this book by examining the growth of German race science from the eighteenth century to the early twentieth and looking at how the Jews were represented by it. The terms *race science, anthropology,* and *medicine* are all used interchangeably because at that time physical anthropologists (another term for race scientists) were generally medical doctors self-taught in the methodology of race science or instructed by other race scientists. Medicine was suffused with the language of race science, and articles in medical journals discussed particular diseases or pathological states as though they were racially determined. Even when not specifically antisemitic, German medicine was implicitly hostile in its constant separation of Jews from other groups in German society.

The treatment of Jews in a nation's anthropological literature directly corresponded to the social position of Jews in that country. Chapter 3 is a case study of the image of the Jew in British anthropology. I argue that the relative insignificance of Jews in that literature, and the particular tone adopted toward them, reflects the relatively propitious circumstances under which English Jewry lived.

Chapter 4 focuses on the first Jewish race scientist, the Anglo-Australian Joseph Jacobs. His work, notable for its broad scope, touched on almost every question concerning contemporary Jewish life in order to determine whether the Jewish condition was determined by nature or nurture. His was the very first attempt to provide empirical, scientific answers and rebuttals to the Jewish racial question.

Jacobs concentrated almost entirely on the condition of European Jewry, and especially the Jews of Western Europe. The lopsidedness of his agenda was corrected by the Russian Jewish race scientist Samuel Weissenberg, who in 1895 began his prolific career studying the Jews of Eastern Europe and the non-Ashkenazic branch of world Jewry. His efforts are the primary focus of chapter 5.

In chapter 6, I investigate the link between science and the politics of Zionism. Zionist physicians used the language of race science to define the Jewish people, defend them against the latest wave of antisemitism, and revive what they regarded as the flagging Jewish identity of German Jews. This group, the most overtly politicized of the Jewish anthropologists, seemed less concerned with the normative methodology of race science, that is, comparative anthropometry (the results of which were often used to point to the superiority or inferiority of certain races) than it was with using the findings of science to effect internal social and attitudinal change among Jews. The Zionist critique of the Diaspora Jew often entailed a damning rebuke of his ghetto physique and mentality; no other contemporary nationalist ideology offered up such an assessment of its own people. But Zionism was predicated upon the belief that only a changed environment could bring about the creation of a new Jewish identity. The work of the Zionist physicians demonstrates how they employed the latest tools of medical and social science, such as population statistics, to bolster a political cause that was concerned first and foremost with Jewish spiritual and physical regeneration.

This does not mean that for Jewish scientists, including Zionists, the idea that there was such a thing as the Jewish race was unimportant. Ignaz Zollschan, for example, constantly stressed the idea of the "racial worth" of the Jews and their contributions to civilization. At the turn of the century, Zionists, together with Europeans representing all political and cultural traditions, believed in the concept of race. Despite great variations of interpretation about the meaning of such biological differences, few questioned the legitimacy of such distinctions. And since the Zionist scientists discussed here were physicians, it is important to point out that medicine actually contributed to spreading the belief in the idea of race by turning race into a medical category.

The distinguished historian George L. Mosse wrote that "a belief in the reality of race did not mean that any one race was necessarily superior to another."[15] He made this remark in reference to Elias Auerbach and other Zionists, some of whom were able to believe in the existence of a Jewish race, even in its purity, without compromising a belief in the fundamental equality of all peoples. The evidence to be brought forth in this work makes Mosse's claim applicable to all Jewish race scientists irrespective of their nationalities or political and religious commitments.

✢ TWO ✢

GERMAN RACE SCIENCE: THE JEW AS ESSENTIAL OTHER

> Do you remember, he asked me, what Lueger, the anti-Semitic mayor of Vienna, once said to the municipality of Vienna when a subsidy for the natural sciences was asked for? "Science? That is what one Jew cribs from another." That is what I say about *Ideengeschichte*, history of ideas.
> *Isaiah Berlin, reflecting on a conversation with Lewis Namier*

During the Enlightenment and the political revolutions that followed in its wake, a fundamental principle of western society was enunciated, that all people are created equal. It is a paradox of Enlightenment thought that just as this principle was given expression, another, potentially contradictory one—that all people are created different—was also put forward. In the eighteenth century, with increasing contact between Europeans and other peoples as a result of slavery, European imperialism, and the great voyages of discovery, the reality of human differences was translated into notions of racial superiority and inferiority.[1] Even the Enlightenment concept of the noble savage, a seeming corrective to the assumption of white superiority, merely reaffirmed extant biological and cultural barriers between whites and non-Europeans and always championed the superiority of white, and especially French, civilization.[2]

The eighteenth century also witnessed the birth of modern anthropology and biology, and the belief in the qualitative difference of races immediately became part of the discipline's methodology.³ Anthropologists were initially concerned with the task of human classification. Taxonomic description, however, was but a short step to the construction of a racial hierarchy. Drawing on the concept of the Great Chain of Being, an idea based on Aristotle's image of the *scala naturae,* anthropologists posited that all living organisms formed a continuous ladder of ascent. In this schema, the bottom rungs were occupied by the simplest organisms, and those on top by the most complex—the divine. Humans were said to occupy a middle position wherein they were linked to God by an infinite number of more perfect celestial bodies. That part of the ladder occupied by humans saw top place go to white European males (closest to God), followed by white women, descending through the various races, with black men and then black women just above the great apes.⁴ In addition, the Enlightenment's challenge to the authority of Christianity and fundamentalist interpretation of Scripture also helped sunder the traditional view of human unity.

The notion of the ascending complexity of organisms was also allied to Enlightenment aesthetic values of human beauty and ugliness. The classical idea of beauty was the measure by which Europeans of the eighteenth and nineteenth centuries determined what was and what was not visually pleasing. Idealized images of beautiful races were created and juxtaposed with stereotypes of racial ugliness. In Germany, the place where physical anthropology can be said to have been born, these ideas were very clearly articulated. One of the most famous expressions of this aesthetic geography was Johann Friedrich Blumenbach's description of the Caucasian as the most handsome of all peoples. It was Blumenbach, the founder of modern anthropology, who coined the term *Caucasian:* "I have taken the name of this variety from Mount Caucasus, both because its neighbourhood, and especially its southern slope, produces the most beautiful race of men, I mean the Georgian."⁵ Christoph Meiners, a professor at Göttingen, declared that "only the Caucasian race deserves the title of beautiful, and with justification, the Mongolian that of ugly."⁶ Predictably, racial Others occupied positions on the scale much closer to the apes than did white, Christian men.

In fact, beauty became quantifiable in the 1770s when the Dutch anatomist Peter Camper invented the facial angle. The measurement was taken by regarding the head in profile and drawing a horizontal line from the nose to the ear and a vertical one from the upper lip to the forehead. The more the jaw protruded, the greater the angle, and thus, according to Camper, the more it resembled the angles of apes and dogs. According to this system, the Negro was more like the ape than like the Caucasian, who tended to have a facial angle of between 70 and 100 degrees, which was said to correspond to the art historian J. J. Winkelmann's "beau idéal," Greek sculpture of the fifth century.[7]

A variation on this theme, but one which still rested on the theory that there existed beautiful and ugly peoples, was the division of people into "active" and "passive" races by Gustav Klemm. In his *General History of Civilization* (1843) Klemm posited that the active races were masculine, thriving in cold climates, while the passive ones, residing in warm climates, were effeminate.[8] In the work of the Detmold anthropologist C. Weerth, who regarded all human differences as the result of environmental influences, the physical beauty of a people reflected its spiritual capacity. After extolling the outward perfection of the Caucasian, Weerth wrote that "this harmonious development of the body corresponds to that of the intellect. Neither in the hot nor the cold zones has a man of greater intellect [than the Caucasian] appeared."[9]

As this kind of classification and description of human beings gave way to the development of comparative anatomy and anthropometry, physical anthropology, or race science, was dominated by quantification and the measurement of various body parts in order to establish the boundaries of human difference. The mathematization of medicine and anthropology was part of a major change that took place within medical science in the 1840s. The scientific school of medicine founded in Germany by Emil Du Bois-Reymond, Ernst Brücke, Hermann Helmholtz, and Carl Ludwig, introduced into medicine a mechanistic model of practice that superseded the older one of describing and classifying diseases.[10]

Yet in spite of the attempt to create a mathematically based—and therefore supposedly objective—theory of human difference, the nascent biological sciences, which included medicine, were not only highly subjective but were often racist. Ironically, it was the mathematization of physical anthropology that assisted in the development of scientific racism and the reaffirmation of preexistent prejudices. For example, numbers that indicated lower brain weights or smaller cranial dimensions were cited as proof that groups possessed of more modest statistics were inferior beings.

In nineteenth- and twentieth-century Europe, medicine and anthropology were saturated with this biologically deterministic language and reductionist methodology. In all countries it was a common assumption in the literature that dealt with human difference that Europeans were racially superior to non-Europeans. This universal discourse, which was concerned with creating biological definitions of Otherness, was, despite methodological and even ideological similarities, not identical the world over. The objects of Otherness in physical anthropology tended to vary from country to country according to such factors as empire, colonial holdings, European politics, ethnic diversity within a given society, and especially the sociological characteristics of that diversity.

In Germany, the Jews constituted the Other in the discourse of scientific racism to an extent not found anywhere else in Europe. While it is tempting to suggest that this has to do with the fact that unlike Britain and France, Germany lacked a substantial colonial empire and thus the nation's scientific attention was turned towards the most accessible national minority in its midst—the Jews—this is not entirely correct. The professional journals of German anthropology are full of articles on African, Asian, Polar, and Pacific cultures. So too were similar scholarly periodicals in other European nations. Where Germany differs is in the extent to which it also discussed the Jews as a race. The reasons for this have to do with the social position of German Jewry. By the turn of the century, the Jews were Germany's most significant minority. Although they only formed about 1 percent of the population (586,833 in 1900), they were highly visible owing to their prominence in the liberal professions and commerce in the cities and as merchants and traders in small towns. Moreover, as societal outsiders, the Jews of Germany, especially in the last third of the nineteenth century, were exposed to an anti-Jewish campaign based on the perceived racial difference of the Jews and the danger that essential difference posed for Germans.

It is no accident that German fears about Jewish difference were heightened just as Germany embarked upon its protracted quest for national unification and state building. The Jews, historically regarded with suspicion, were now seen as a threat to the nation. The kind of threat the Jews posed varied according to the ideologies of the different power groups in society. Nevertheless, all were able to pin their frustrations on the Jew. To nationalists, Jews still formed a state within a state and therefore were a potentially traitorous entity. To groups touting a völkisch ideology, lamenting the rapid and unprecedented changes that Germany was undergoing due to unification and industrialization, the Jews were charged with being responsible for and in the forefront of those changes. To many on the left, the Jew represented the interests of capital and therefore, exploitation. Conversely, the Jews were often denounced as the bearers of socialism, and promoters of revolution.

This perception of the Jew as outsider made its way into the literature of the biological sciences. Although they classified Jews as members of the white race, German medicine and anthropology also isolated Jews and referred to them as a group apart, thus reflecting the Jews' dual position of being German but not being fully part of German society. The scientific literature reveals an uneasiness and discomfort with Jews. The German medical and anthropological communities were unable to come to terms with Jews within the paradigms that German society set for minority group integration. Nearly one hundred years after the Berlin Jewish philosopher Moses Mendelssohn attempted to reconcile traditional Judaism with *Deutschtum,* thereby establishing a model for future Jewish integration into the modern state, German Jews of the late nineteenth century remained a group apart. Voluntarily, social integration had remained incomplete. And although German Jews enthusiastically partook of and contributed to German culture, it is also the case that German Jews creatively developed a unique Jewish subculture designed to coexist with and complement German culture.[11] Long after emancipation, then, the Jews continued to exist as Jews. The overwhelming majority did not forsake their religion, ethnic pride in their Jewishness, or even their affiliations to the local Jewish Gemeinde, the officially sanctioned Jewish bodies regulating communal and ritual affairs. Every Jew was required by law to hold membership in the Gemeinde nearest his or her place of residence. Secession from the Gemeinde was only possible after filing a petition with state authorities asking permission to be relieved of membership. The hopes of non-Jewish proponents of Jewish emancipation such as the eighteenth-century Prussian bureaucrat Christian Wilhelm Dohm, that if treated humanely the Jews would eventually convert to Christianity, thus remained unfulfilled.

Nevertheless, German Jews felt that by adopting German language and culture, they had fulfilled their part of the emancipation "contract." Gentiles, however, found it perplexing and frustrating to see Jews make such strides towards becoming German and yet seem unable to leave their Jewishness behind. This situation is reflected in the many paradoxes of Jewish life in Germany. Movement towards professional integration was met by a countermovement towards social isolation; the high visibility of Jews contrasted with their demographic marginality; and the physical and linguistic indistinctiveness of Jews from gentiles belied a drive towards Jewish ethnic independence. It was the elements of this paradox that combined to form the so-called *Judenfrage* or Jewish question. The medical profession echoed in the language of science the Germans' concerns about the anomalous position of the Jews. German medicine and anthropology lent their prestige and authority to the debate, attempting to provide scientific evidence to determine whether biological or racial peculiarities were the cause of the Jewish question.

Although Jews were not always represented in this literature in an explicitly antisemitic light, they were almost always spoken of in terms and tones that accentuated their uniqueness. Alfred Ploetz, for example, the founder of the German racial hygiene movement (1905), wrote in 1895 that the Jews, whom he regarded as "more Aryan than non-Aryan," were, together with the West Aryans, the most highly developed of the so-called *Culturrassen* or civilized races. Because of the complete physical and mental compatibility of West Aryans and Jews, Ploetz deplored that an "artificial wall" had been established between the two groups and advocated instead the "complete absorption" of the Jews into the West Aryan race. This, Ploetz believed, was not only in the interest of the middle class but would also bring about the "refinement" of both groups.[12]

As a socialist, Ploetz fits into that left-wing tradition of European thinking dating from the French revolution that denied the Jews their right to exist in society as Jews. The most famous example of this thinking was expressed in 1789 by Count Stanislas de Clermont-Tonnerre, a revolutionary deputy to the National Assembly and an advocate of equal rights for the Jews. He declared in the chamber during the debate on the eligibility of the Jews for citizenship that "the Jews should be denied everything as a nation, but granted everything as individuals." That is, the Jews should be granted citizenship, but they must give up claims to national, communal, and judicial separateness.[13] In keeping with this tradition, Ploetz also paid lip service to his detestation and contempt for antisemitism, maintaining that it would be swept away in the flood tide of scientific knowledge and humane democracy.[14] Nonetheless, it is no accident that he chose Jesus, Spinoza, and Marx as the three most illustrious representatives of the Jews. No doubt for Ploetz they were perfect models of the "completely absorbed" Jew.

The implicitly antisemitic image of Jewish separateness and essential difference in German anthropology prior to 1918 was exemplified in Ludwig Woltmann's *Politisch-Anthropologische Revue*. From its founding in 1902, this journal remained for two decades the central organ for the propagation of the idea of Aryan superiority. The Jewish anthropologist Ignaz Zollschan railed against Woltmann, drawing the obvious conclusion that a theory granting only to the blond dolichocephalic peoples of the world unrivaled genius, courage, and virtue was exclusionary and therefore racist. According to the *Revue,* the qualities of blacks, Asians, and Semites were diametrically opposed to those of the Nordics.[15] But in the social reality that existed before World War I, the only significant contact Germans had with any of these three "inferior" groups was with the Jews, represented in Woltmann's schema by the Semites. After the war the *Revue*, which changed its name to the *Politisch-Anthropologische Monatsschrift*, became explicitly antisemitic.[16]

Other scientists who held similar views were Eugen Fischer and Fritz Lenz, who eventually became the leading race scientists of the Third Reich. Before 1933, however, they and other members of the German racial hygiene movement were primarily concerned with preserving the Aryan race, which they believed was threatened by "counterselective" forces such as war, revolution, and improved medical care. According to this theory, war and revolution served to kill off the fittest, while advanced medical technology enabled the unhealthy to survive. The Jews were not singled out as being one of those forces posing a threat to the Germans. In its formative years the German racial hygiene movement was simply a part of the international eugenics movement.[17] In fact, significant contributions were made by Jewish scientists to the movement's internationally renowned journal, the *Archiv für Rassen- und Gesellschaftsbiologie*, and members of the German racial hygiene movement even trained Jewish physicians. For example, at the Leo Baeck Institute in New York, the file of a German Jewish émigré doctor and race scientist from Berlin, William Nussbaum, contains a diploma showing that in 1931 he had successfully passed his course of study at the Kaiser Wilhelm Institut für Anthropologie, menschliche Erblehre und Eugenik. Fischer, who taught at the institute, personally signed Nussbaum's certificate.[18] Yet despite Fischer's willingness to train Jewish students, he still wrote of Jews as though they were genetically different from the Germans. In 1914, for example, in *The Problem of Racial Crossing among Humans,* Fischer posited, on the basis of Mendelian genetics, that certain racial characteristics, such as the "Jewish" nose, were recessive. For Fritz Lenz, Jewishness was indelible. In *Outline of Human Genetics and Racial Hygiene* (1927) he wrote: "Jews do not transform themselves into Germans by writing books on Goethe."[19]

The Jewish racial question focused upon determining the physical and psychological characteristics of the Jews in order to ascertain whether these differed fundamentally from those of non-Jews. The debate over the physical nature of Jewishness revolved around determining whether the Jews were composed of one or more racial types, and whether they displayed pathological features peculiar to them as Jews. The discussion about the psychological nature of Jewishness centered upon the classification and etiology of mental illnesses suffered by Jews, and once again, whether these were quantitatively or qualitatively different from the psychological disorders of non-Jews. In both the anthropological and psychological sense, the German natural sciences cast the Jews into the role of the pathological Other.

The question of Jewish racial purity was the most hotly debated issue among race scientists who studied the anthropology of the Jews. Until the end of the nineteenth century, German anthropology unanimously regarded the Jews as racially pure. Influenced by the social and historical reality of Jewish separateness, anthropologists assumed a priori that Jews were a discrete and identifiable biological entity. German anthropologists rarely began a discussion about Jews without noting that for the last two thousand years the Jews had refrained from mixing with the peoples among whom they lived, having thus retained their original racial characteristics. Many race scientists, even if they were otherwise antisemitic, lauded the so-called racial purity of the Jews, for it bespoke an enviable racial awareness and instinct for survival.

From the earliest days of German physical anthropology, the Jews were classified as a single uniform type, easily identifiable, readily distinguishable from their neighbors, and impervious to the effects of the environment. In the 1795 edition of his classic work *De generis humani varietate nativa*, Johann Friedrich Blumenbach treated the Jews as an exception to the rules of nature that he himself posited concerning racial variation. Blumenbach held that differences in human appearance were conditioned by climate, and those characteristics were susceptible to alteration when a people migrated to a new place. The Jews, however, did not seem to fit this rule. Wide geographic dispersion and various climatic conditions were unable to effect a change in the Jewish countenance: "Above all, the nation of the Jews, who under every climate remain the same as far as the fundamental configuration of face goes, [is] remarkable for a racial character almost universal, which can be distinguished at the first glance even by those little skilled in physiognomy, although it is difficult to limit and express by words."[20] Blumenbach's own inability to describe that which existed only in his imagination speaks volumes about scientific racism and its use of the crutch of racial myths to prop up arguments and observations that defied empirical investigation.

A Dutch anatomist by the name of Wachter published his study of the skull of a thirty-year-old Jewish man in 1812. Familiar with the work of Blumenbach, Wachter was even more specific, noting the peculiarly "large nasal bones," the "square chin," and the specifically Jewish "bony impressions on both sides of the lateral orbits." The cause of these impressions was that "among Jews, the muscles primarily used for talking and laughing are of a kind entirely different from those of Christians."[21]

Also in 1812, the year the Jews were first emancipated in Prussia, the Berlin anthropologist Karl Asmund Rudolphi remarked on the specificity of Jewish physical characteristics. Rudolphi observed that the physical differences of Jews from Germans reflected their social isolation. Strained social relations, according to Rudolphi, seemed to echo the imperfect physical harmony that existed between Jews and Germans. The physiological difference of the Jews was the cause of their social separateness. Curiously enough, according to Rudolphi, the Jews' long tenure in European climates had not altered their outward appearance as Jews, that is, they remained uniquely distinguishable from the majority population: "Under Julius Caesar [the Jews] were almost as deeply rooted in Rome as they are today in some states of Germany and in Poland and, in a word, have become indigenous.... [But] their form [*Gestalt*] has not changed. Their color is here lighter, there darker, but their face, their skulls everywhere have a peculiar character."[22] Of peoples in general, Rudolphi concluded: "Some remain in their homeland such as blacks and Papuans, others emigrate such as the Jews, but all, so long as they remain unmixed, bear the unmistakable signs of their individuality."[23]

The view in Germany that the Jews formed a single, homogeneous race was remarkably tenacious. In 1881, almost a century after Blumenbach, the renowned geographer and ethnologist Richard Andree published *Zur Volkskunde der Juden,* one of the first books of the modern era devoted to the anthropology of the Jews. It was Andree's opinion that the Jew was easily and everywhere recognizable and, moreover, that "anthropologically, the Jews are the most interesting objects, for no other racial type can be traced back through the millennia with such certainty as the Jews. And no other racial type displays such a constancy of form, withstanding the influences of time and environment as does this one. They have overcome proportionately strong admixtures of foreign blood, and ... no new type, no new amalgamation has taken place. The Semitic blood carried in the most decisive way the victory [for the Jews] and the ancient monumental Jewish body remains preserved, as does the ancient hereditary Jewish spirit."[24]

Andree's assertion of the unique spiritual character of the Jews introduced a new and important element into the now familiar argument of Jewish identifiability. It must be recalled that Andree was writing at a time when the mass of Jews fleeing pogroms in Russia had begun to enter

Germany and the antisemitic movement had started to snowball. The backlash against these foreign Jews was directed with equal ferocity against those "assimilated" German Jews who had, since the time of Moses Mendelssohn, sought to shed their peculiarly Jewish characteristics and blend into German society. The presence of East European Jews in Germany seemed to set that process back, reminding indigenous German Jews of their own cultural heritage and confirming to the oppositional voices among the Germans their a priori attitudes concerning the essence and unchangeability of the Jew.

Scientists like Andree propounded a controlling discourse which attempted to identify, then divide and conquer all Jews, by separating them from Germans and denying them the right and ability to assimilate. Andree was quite explicit about this. After a summary of the most recent anthropometric research on Jews, he wrote: "We all recognize the Jewish type. We immediately distinguish him by his face, his habits, the way he holds his head, his gesticulations or when he opens his mouth and begins to speak. And it is always possible . . . to recognize even the most assimilated, because he always bears some characteristic of his race."[25]

Andree was speaking here of "the Jew," and his description was therefore stereotypical. He claimed that all Jews appeared and acted the same, and in this respect his anthropological methodology was somewhat dated. Already in the 1860s, under the influence of the polygenists and the sway of rising nationalism, anthropology split peoples into myriad groups and types, eschewing the more static and uniform pictures of nations it had previously painted. This new development posed a challenge to the theory of Jewish racial uniformity. Focusing on head shape as an indicator of relative purity because the head was held to be the most stable racial feature, scientists were at pains to explain how Jews, if they were Semites and a pure race, no longer resembled either ancient Semitic types or modern Arabs in craniometric dimensions. New scientific techniques were employed to help solve the dilemma. Comparative craniometry (the measuring of skulls) was the defining methodology of modern race science. Its origins date from 1844, when the Swedish anatomist Andreas Retzius introduced the cranial index. This became the crucial measure of race. To calculate the index, anthropologists multiplied the width of the head by 100 and divided the product by the length. A cephalic index above 80 was regarded as brachycephalic; between 75 and 80, mesocephalic; and below 75, dolichocephalic.[26]

The earliest systematic exposition of the theory that the Jews displayed two distinct cranial types, round brachycephalic and long dolichocephalic, was by the naturalist and physician Carl Vogt. In 1864 Vogt published his celebrated *Lectures on Man,* in which he clearly asserted, for the first time, that not all Jews looked alike and that there were two distinct yet related branches of Jewry: "[There exists] chiefly in the North, in Russia and Poland, Germany and Bohemia, a tribe of Jews frequently with red hair, short beard, pug nose, small grey cunning eyes, massive trunk, round face and broad cheek bones, resembling many Sclavonian tribes of the North. In the East, on the contrary, and about the Mediterranean, as well as in Portugal and in Holland, we find the Semitic stock with long black hair and beard, large almond-shaped eyes with a melancholy expression, oval face and prominent nose; in short, that type represented in the portraits of Rembrandt."[27]

In distinguishing one group of Jews from another, Vogt's observations mark a sharp change in the direction of German anthropology's discourse on the Jews. No longer represented as a homogeneous mass, the Jews were now recognized to be composed of two distinct racial elements. Yet Vogt's conception still complied with tradition in two central ways. First, he dismissed the notion that climate could effect bodily change. According to him, irrespective of where Jews currently lived, they had not undergone any morphological transformation due to the environment but rather displayed original and unchanging racial traits. Second, Vogt's description of Jews betrays his reliance on traditionally established notions of beauty and ugliness. Nevertheless, although the two sorts of Jews looked different from each other, they still remained recognizable enough to distinguish them from non-Jews. In these two respects, Vogt's pronouncements on the Jews were really nothing new. Craniometric examinations notwithstanding, the scientific evidence introduced by Vogt merely confirmed the long-established belief in Jewish racial difference.

After Vogt, the view of the anthropological "splitters" held sway. This did not, however, mean complete consensus on the issue. For example, Vogt ascribed the East European Jewish type to the admixture of Slavic blood, and four years later Friedrich Maurer, who concurred with Vogt that the Jews presented two racial types, posited that the racial split was caused by the introduction of Turanian elements.[28] The Austrian physician Augustin Weisbach published the findings of his craniometric examinations of nineteen Jewish males in 1877, and concluded that "among the European Jews there are doubtless two cephalic types. One is dolichocephalic, with a narrow, long face, on the whole a big nose, and thin lips; and the other brachycephalic, with a wide face, low, broad, small nose, and thick lips."[29] Similarly, Bernhard Blechmann's study of 1882 convinced him that present-day East European Jews were racially different from their Semitic forefathers.[30]

Two other German anthropologists whose views were widely regarded by contemporaries, Ludwig Stieda, professor of anatomy at Dorpat, and Julius Kollmann, professor of anatomy at Basel, both maintained on the basis of cranial measurements that the Jews were composed of two racial types, and that the brachycephalic Ashkenazic was completely different from the original dolichocephalic Sephardic.[31] Kollmann found that the twelve medieval skulls he studied in 1885, from the ancient Jewish cemetery in Basel, had an almost hyperbrachycephalic cranial index of 84.74.[32] For Kollmann, this was clear proof that the Jews were originally composed of two racially distinct types. He was unable to entertain any other explanation because of his unshakeable belief in the permanency of racial characteristics and in Jewish racial purity.[33] Just how such morphological divergences arose among the Jews, including hair, eye, and skin color as well as widely differing bodily measurements, became the central debate in modern anthropological theories about the Jews.

Scientists were not content to split the Jews anthropologically in two. Consistent with larger trends in anthropology, which divided Europeans into countless racial varieties, Jews were said by some anthropologists to consist of three racial types. The view of these scientists, most clearly represented by Constantine Ikow, was that only the Semitic type was original and the two others were products of the Jewish Diaspora. Ikow divided the Jews into the following groups: (1) those from the East and South, the Balkans, Turkey, Algeria, Spain, Italy, and around the Mediterranean; (2) those from Western Europe, who had intermixed with the native populations; and (3) Russian Jews, who show no Semitic characteristics at all.[34] Ikow's theory was radically reinterpreted by the French anthropologist Gustave Lagneau, who in 1891 noted that only among the Jews of North Africa were there traces of Semitic blood. According to Lagneau, the Jews of the Mediterranean had mixed with Greek and Roman proselytes, while Jews from Eastern Europe were composed of Germanic and Slavic elements.[35] The debate over the extent to which proselytism and intermarriage influenced the physical appearance of the Jews, and whether these two phenomena were responsible for the racial splits among them, was another issue of central concern to race science in general and Jewish race scientists in particular.

Not all anthropologists were so keen to classify Jews. By the last decade of the nineteenth century, some had begun to deny the Jews were a race at all, preferring to see them as a community of faith, a *Religionsgemein-*

schaft. According to many of the German scientists who put forth this view, the Jews were a mixed race, showing no uniformity of appearance. Such was the view of the great pathologist Rudolf Virchow. In 1886 he conducted a massive anthropological survey of more than ten thousand schoolchildren to determine the racial makeup of Germany. He demonstrated that not only were at least 10 percent of Jews blond but that, contrary to popular myth, only 31 percent of Germans were blond, thus proving conclusively that there were no pure races in Germany.[36] Although Virchow conducted a survey that separated Jewish from Christian schoolchildren, he must not be seen as antisemitic. Virchow's discovery of cellular pathology, his work in epidemiology, anthropology, and public health, his position as a leading member of the Progressives in the Prussian and German diets, and his battles against the antisemites fully entitle him to George Mosse's epitaph that he was "one of the last universal scientists our civilization has produced."[37] Nevertheless, his survey underscored the gulf that separated Germans and Jews. By classifying the two groups separately, Virchow tacitly acknowledged that in the minds of Germans, emancipation had failed to eradicate Jewish differences. The Jews still remained identifiable and distinct, even to someone such as Virchow who bore them no ill will. The Virchow survey is an excellent example of the negative attitude toward Jews that was implicit in German medicine of the nineteenth century. The very nature of the survey and its methodology reflected the profession's and, by extension, society's refusal to accord the Jew completely equal status. The authority of medicine was used to delineate racial differences in order to fortify social barriers that were being gradually eroded by the breakdown of the old order and the onset of modern society.

For Virchow, a political liberal, the findings of his survey confirmed that Jewish-Christian integration was progressing and progressive.[38] For him, the blending of races (the assumption being that at one time in the past pure races did exist) was a positive development. Like the Germans, the Jews displayed a variety of racial characteristics, and during the course of centuries of relative seclusion, original differences among the various Jewish tribes (*Stämme*) became more pronounced. At the same time, certain foreign admixtures lead to the creation of an average "Jewish national type (*Nationaltypus*)." According to Virchow, racial differences were merely a matter of hereditary variations. They were not unbridgeable divisions. Although the crossing with Germanic racial elements in the Jewish Diaspora accounted for Jewish blue-eyed blonds, the "Jewish national type" differed, of course, from the average "German national type." It must be stressed that Virchow saw no qualitative differences stemming from such a typological variation. He steadfastly refused to extrapolate from the statistics that physical differences could translate into moral or intellectual ones.

Not all anthropologists agreed with Virchow concerning his timetable for the mixing of Jews with other peoples. A very influential school of anthropological thought developed in late nineteenth-century Germany that ascribed the light pigmentation of Jews to their prehistoric intermixture with Aryans. Denying that the Jews were a pure race, Moritz Alsberg noted that "even before the Israelites had taken Palestine as their land, a part of this country was occupied by a race whose characteristic peculiarities were those of the Germanic branch of the great Aryan race."[39] From this branch the Israelites had received their Nordic features. The celebrated professor of anthropology at the University of Berlin, Felix von Luschan, also denied that the Jews were ever a pure race, noting that their brachycephalism was a result of having early on mixed with roundheaded Hittites.[40]

The question for anthropology was, At what point in their long history had Jews mixed with non-Jews to produce racial divergencies from the Semitic norm? Had racial mixing occurred in ancient Palestine, so that the Jews were not a pure race even in the time of their political independence, or had it taken place during their long period in exile? Drawing on the historical record, Jewish anthropologists fought well-publicized battles with non-Jewish scientists over this very point.

Whether they were regarded as a pure, bifurcated, or mixed race, or even as a religious community, German fin-de-siècle medical and anthropological discourse regarded Jews as a group apart and interpreted that separateness in qualitative terms. It is not surprising that a direct corollary of the theory of Jewish anthropological uniqueness was the belief that the Jews also possessed a unique racial pathology. Medical science postulated that Jews were racially protected from certain contagious diseases and at the same time prone to unique psychopathologies.[41]

An example is alcoholism, which was first viewed as a disease in the nineteenth century. Accepting as a given that there was such a thing as a particular racial pathology, scientists studied the drinking habits of many peoples in order to compare, contrast, and rank the races on a scale of sobriety. Popular wisdom had it that Jews rarely drank and that they had barely any alcoholics among them. Researchers reasoned that if this were an environmentally determined problem, then Jews in Poland or Russia should have displayed similar proportions of drunkards among them as

Poles and Russians. Since they did not, nearly all the findings of gentile specialists on the subject attributed Jewish temperance to racial factors, confusing the culture of Jewish moderation in drinking with racial immunity or a hereditary aversion to liquor.

Similarly, ancient superstitions that Jews did not succumb to plague and pestilence to the same extent that non-Jews did were widely believed to hold true in modern times with regard to typhus, tuberculosis, and cholera. Childhood illnesses such as measles, scarlet fever, diphtheria, and croup were said to have found fewer victims among Jews than non-Jews. Above all, Jews were said to enjoy greater longevity and lower infant mortality than non-Jews. On the other hand, Jews were believed to suffer inordinately from diabetes, known in German as the *Judenkrankheit*, respiratory ailments, hemorrhoids, cancer (but neither penile nor uterine, due to male circumcision), conjunctivitis trachoma, and color blindness.[42]

Fin-de-siècle medicine also maintained, with near unanimity, that Jews, because of their race, suffered from a higher incidence of insanity than did non-Jews. Two of the most influential psychiatrists of the nineteenth century, Emil Kraepelin and Richard von Krafft-Ebing, expressed mainstream psychiatry's evaluation of the Jews. In *Psychiatrie: Ein Lehrbuch*, which was first published in 1883 and went through nine editions by 1927, Kraepelin noted that it was impossible to tell which of the various influences, race, lifelong habits, climate, diet, or general health conditions, were responsible for insanity. But with the Jews, Kraepelin was sure that race did play an etiological factor in their mental illness. "That the peculiarity of a people," he wrote, "can play a role [in causing insanity] is proven by the case of the Jews, who can be compared to the surrounding population without making great [statistical] errors. The comparison shows that at least in Germany, and likewise in England, the Jews are inclined to mental and nervous disease in considerably higher numbers than the Teutons. Certainly alcoholic forms of insanity are rare amongst them; on the other hand we see such extraordinary disturbances caused by hereditary degeneration."[43] Krafft-Ebing was inclined to regard the higher frequency of mental illness among Jews as a consequence of their religious fervor, which in turn encouraged deviant sexual practices such as consanguineous marriage: "Very often excessive religious inclination is itself a symptom of an originally abnormal character or actual disease, and, not infrequently, concealed under a veil of religious enthusiasm there is abnormally intensified sensuality and sexual excitement that lead to sexual errors that are of etiological significance."[44]

To charge German Jews with "excessive religious inclination," or to describe Judaism as it was practiced in fin-de-siècle Germany as religiously channeled sensuality, was to be woefully ignorant of the reality. No account is made for the various modes of Jewish religious expression or for that uniquely modern creature, the "godless Jew." Judaism is spoken of as a homogeneous religious culture and all Jews as identical to one another—passionate and pious. For Krafft-Ebing, to be Jewish was to be afflicted with religious fanaticism and "psychopathia sexualis."

German psychiatry's marginalization of the Jews completed the attempt of the medical profession to represent them as a biologically separate group. Whether it portrayed them as physically different—and this could even mean beautiful, especially when speaking of Sephardic women—or psychologically distinct from Germans, whose supposed contemplative nature was more apt to result in chronic melancholia than in the hysteria of the hot-blooded Semitic Jew, German medicine sought to return Jews to a theoretical ghetto of biological difference.[45] It is no accident that this process occurred just at the time when socially and especially culturally, barriers between the two communities were at their lowest.

Emphasizing the differences ensured the Jews' identifiability. Their post-emancipatory integration into German society, no matter how incomplete, was sufficiently well advanced in the cities to blur obvious and ancient distinctions that, after German unification in 1871, needed reclarification in the wake of changed political and social realities. The Germans embarked upon a program of state building, national integration, and consciousness raising. Nationalism and racism, both scientific and popular, were important tools in this construction project. In the newly unified Kaiserreich, national integration meant the incorporation of similar peoples, and the exclusion of traditional Others. By the turn of the century, the most conspicuous Other in Germany, the Jews, began to experience an increasingly hostile climate as German nationalist and colonialist aspirations grew in tandem with the state's ever-increasing power. A natural consequence of this situation, in which Germany began to resemble a singular nation-state, was the increased emphasis and reflection upon what it meant to be a German. Myriad antisemitic parties, groups, and organizations, many of whom were affiliated with or had access to the highest levels of German government and society, dealt with this most pressing concern.[46] It was clear to all these people that to be German meant to be Christian, and that to be a German Christian was not merely a national-religious definition but, in some quasi-mystical way, a racial one as well.[47]

Medical science came to the aid of those who posed questions about the essence of German national identity. With its authoritative language and social prestige, it confirmed widespread notions of Jewish racial difference, which in turn reinforced ideas of German homogeneity. But more than this, it expanded the terms of the debate over national identity by providing a scientific discourse that was perceived as unassailable and whose logic, because it was based on statistics, was regarded as unimpeachable.

Jewish scientists, like their German counterparts, used the language and methodology of race science to craft their own explanations for the distinctions between gentiles and Jews. But race science also provided them with a liberating discourse. In the wake of the perceived failures of emancipation and assimilation, anthropology became an ideological tool to free Jews from the humiliation brought on by the loss of Jewish identity. Race science was, in this context, race-affirming. In the hands of Jewish doctors, it was as much meant for internal Jewish, as it was for external, non-Jewish consumption.

The Jewish racial question, as interpreted by anthropologists, centered on the issue of nature versus nurture. The intellectual agenda of Jewish race scientists was to wrest control of the anthropological discourse on Jews from gentiles, in part to ascertain whether the historical experience of the Jewish people, or the inheritance of certain physical and psychological characteristics, determined their present biological and sociological status. Like non-Jewish anthropologists, Jewish scientists sought to determine whether the Jewish people were homogeneous or composed of various racial types. They, too, vigorously debated among themselves the issue of Jewish racial purity. Jewish doctors attempted to define whether Jews were susceptible or immune to certain diseases and displayed other physiological peculiarities. For example, Jewish scientists studied the age at which Jewish girls reached puberty to see if it differed from the pattern among non-Jews. They asked whether Jews lived longer than non-Jews, whether they had more children than non-Jews, and whether marriages between Jews and gentiles were as fertile as those between Jews.

These racial questions were not only restricted to discussions about the Jews' body. They also had important social dimensions which made the Jewish question such a complex one. Were the occupational choices of Jews determined by an innate aversion to manual labor? Was that aversion a result of the universally recognized smaller stature of Jews? Similarly, did those physical impediments disqualify Jews from regular army service and thus from full integration into the modern European nation-state? Were Jews an assimilable group or would they forever remain outside society's mainstream because of certain racial characteristics? At a time of massive westward migration of Russian Jews, this was a particularly pressing issue for governments and Jewish communities alike.

In Germany, the Jews were especially well placed to undertake such a project given their predominance in the medical profession. With few exceptions most race scientists, Jews included, were medical doctors. Not only was medical training the preferred route by which most physical anthropologists learned their craft, but for Central European Jews, their particular experience in the broad culture of German medicine contributed to the creation of a Jewish race science. The proportion of Jews attending university in Germany was far greater than that of the majority population. Although they were only approximately 1 percent of the total population from the mid-1880s to 1933, the percentage of Jews receiving a university education fluctuated between 4.5 and 8 percent. From 1891 to 1911 the number of Jews enrolled in medical faculties remained in a startlingly high range, 10–16 percent.[48]

Within the medical sciences Jews concentrated in clinical rather than academic fields, for two reasons. First, after 1890 clinical medicine underwent a general expansion that created room for professional advancement, and Jews flocked to the specialties of internal medicine, gynecology, ophthalmology, dermatology, and psychiatry. Second, specializing in one area of clinical medicine afforded the young Jewish doctor the opportunity to enter into private practice, whereas research medicine could only lead to a university position. And when Jews, especially unbaptized ones, accepted such appointments, they were usually relegated to the lower rungs of the academic ladder. Furthermore, the theoretical fields were the older ones, with a more entrenched and conservative faculty inclined to deny access and professional advancement to Jews.[49]

A clear example of the bias that Jewish medical students encountered in Central Europe occurred at the Vienna Medical School, in the conservative department of surgery. In 1875, the distinguished professor of surgery Theodor Billroth wrote a seven-page diatribe against the university's open-admission policy for citizens of the state. Although he was primarily referring to East European Jews, he made no clear distinction between them and other Jews. His was a classic attack on all Jews on the basis of perceived racial differences. Billroth forthrightly asserted that "a Jew—just as little as a Persian, Frenchman, New Zealander or African—can never become a German." The Jews of Germany might speak German and be educated in Germany, but they retain their "national tradition"—Jewish—as strongly as Germans living abroad keep theirs. That the Jews might ever become German nationalists was, to Billroth's thinking, neither "expected nor desirable."[50]

This was just one of an endless string of attacks on Jewish doctors.[51] In 1899, the publication of the Centralverein deutscher Staatsbürger jüdischen Glaubens, *Im deutschen Reich,* carried as its lead story "Antisemitism and Medicine." The nineteen-page article detailed expressions of antisemitism within the medical establishment. According to the article's author, Joseph Lewy, Jewish doctors, and even philosemitic non-Jewish physicians such as Rudolf Virchow and Hermann Nothnagel, were repeatedly singled out for opprobrium by the antisemitic press in Vienna. In a modern variant of the medieval accusation of well poisoning, the press had sensationalized a case that concerned a Jewish doctor (and his less well publicized Christian department head) who had purposely infected hospital patients with bacteria as part of their medical research. The author of the *Im deutschen Reich* article deplored the practice but reminded readers that there were also non-Jewish doctors who were guilty of the misconduct, as recent incidents in Leipzig and Berlin showed. Joseph Lewy also pointed out that in Berlin, professional antisemites and members of the medical profession had made the Jewish doctor the "whipping-boy onto whom all blame was shifted when some kind of malpractice somewhere was perpetrated."[52]

The drive among Jewish scientists to establish a Jewish racial identity on a scientific basis was in part motivated by their institutional rejection at the hands of the German medical establishment. Although Jews accounted for a staggering 16 percent of all doctors in Germany in 1900,[53] they were denied prestigious academic appointments, were largely excluded from research positions in theoretical medicine, and were frequently charged with malpractice. So widely publicized was the ostracism and antisemitism encountered by Jewish doctors in fin-de-siècle Germany and Austria that the topic was the central theme of the Viennese Jewish playwright and physician Arthur Schnitzler's *Dr. Bernhardi* (1912).

Specific institutional problems fostered the antisemitic backlash in the medical profession. Lewy identified three factors: intense competition among private practitioners for patients, because recent sickness-insurance laws had severely restricted the pool of patients who had access to private care; an "overproduction of doctors from year to year"; and the fact that "quackery was in full bloom and increasing due to the absence of strict regulations."[54] According to Lewy, these three developments caused the financial difficulties of many German doctors, as well as the worsening of the image of doctors in the eyes of patients and the subsequent loss of professional prestige. The frustrations of many German doctors were vented in the form of antisemitism, expressed either grossly or, within the anthropological-medical debates over the Jewish question, quite subtly.

Thus, Jewish race scientists were also motivated by their own personal humiliation. The use of medical language to assert Jewish equality and, very often, moral superiority (in the pages of strictly professional journals intended only for members of the guilds) clearly indicates that the approach to the Jewish racial question by Jewish physicians also had a parochial subtext. The sociology of medical practice in the Second Empire as it pertained to Jews contains aspects of the larger Jewish question, and reveals that the struggle of Jewish and gentile scientists betrayed the personal and professional motivations of the two competing groups. The way Jews were treated professionally, and represented in the literature of the natural sciences, reflected their position in German society as a whole.

❋ THREE ❋

BRITISH RACE SCIENCE: THE JEW AS NON-OTHER

> The most certain test by which we judge whether a country is really free is the amount of security enjoyed by minorities.
> *Lord Acton,* The History of Freedom in Antiquity

Prior to the closing decades of the nineteenth century, Jews were not regarded as legitimate subjects for anthropological inquiry in England. After the heated debate surrounding the failed Jew Bill of 1753, which sought to give foreign-born Jews the same privileges as their native-born children enjoyed, the Jews eventually became a relatively unobtrusive minority, comfortably ensconced and suffering very few social or political disabilities. This singular quality of being "at home" was echoed in the literature of anthropological science; Jews as Jews simply failed to arouse British scientific curiosity.

It need hardly be stressed, however, that Jews had long been recognized in England as an undesirable element. The first sustained period of Jewish demographic growth in England was in the twelfth century, when dozens of Jewish communities were established throughout the country, the most important in London. The bulk of Jews were engaged in commercial activities, especially moneylending. As a result of their cultural difference from the English, the royal protection they enjoyed, and their lending at interest, the Jews incurred the increasing animosity of both the masses and the rebellious baronage. The first recorded blood libel in medieval

Europe took place at Norwich in 1144, and this was followed by a string of similar accusations over the course of the century.

The crown derived much of its revenue from the profits of Jewish moneylenders, mercilessly taxing and financially bleeding them through the imposition of extortionate tallages. When Edward I ascended the throne in 1272, he found the community destitute, its value to the royal treasury negligible. Moreover, foreign bankers had taken over the credit activities formerly administered by Jews. Once the Jews were unable to perform their traditional services for the sovereign, Edward elected to expel them from England in 1290. With some exceptions, England was almost completely free of Jews until the middle of the seventeenth century, when the revolution and the extreme Puritan atmosphere that prevailed led to a more favorable attitude toward the Jews and to debate over formally readmitting them to England. Although no such order was ever passed, Jews began to arrive and the community continued to increase in size and become more prosperous over the next two centuries. When debates over Jewish emancipation began in 1830, there still were restrictions on Jews opening shops in London, receiving a university degree, or sitting in Parliament. But the impact of these humiliations on English Jewry should not be overestimated, nor should they be seen as unique. Many discriminatory restrictions had also been in place against Catholics, Quakers, and Baptists. In Prussia, prohibitions against Jews becoming reserve officers, members of the upper state bureaucracy, or full professors were of far greater significance and more keenly felt by community members there.

If England can be said to have had an antisemitic tradition, it is most clear in the image of the Jew in English literature. From Chaucer's *Prioress's Tale* and Shakespeare's Shylock to Rebecca in Walter Scott's *Ivanhoe* and Dickens's Fagin, the Jew had been presented as an alien figure consumed by vengefulness, miserliness, and avarice.[1] By the nineteenth century, however, the stereotype no longer reflected England's treatment of the Jews, who had become well integrated into English society and, after the removal of the last remaining disabilities, never again came under serious threat as a community. It is social reality, not literary convention, that determined the image of the Jew in Britain, an image most closely reflected in the texts of anthropology and ethnography. In those works the Jews are mostly benign and unobtrusive, and are relegated to the periphery—a mirror image of their demeanor in English society and culture.

Peoples who inhabited some distant corner of the British empire, who were "primitive" and in need of "civilizing"—the "lower" races, invariably non-Caucasian—were the subject of British anthropological research.² British Jewry did not fit the description. From the eighteenth century on, Continental Jews who settled in England, rich and poor alike, eased comfortably into British society. Most resided in London, and some of those in the rapidly expanding bourgeoisie earned their livelihoods in the City. Even lower-class Jews, the majority of whom were involved in petty trade, tended to acculturate to the general patterns of English working-class life without too much trouble.³ Thus, up until the great influx of East European Jews in the 1880s, the very Englishness of Anglo-Jewry did not provide the stuff that great anthropological debates are made from.

This failure to arouse the interest of British anthropologists tells us as much about the successful integration of Jews into English society as it does about the concerns of the nascent discipline itself. There seems to be a direct correlation between the degree to which Jews were addressed in anthropological literature and the social position their community enjoyed in any given European country. In Germany, the problem of whether the Jews could ever become Germans and participate fully in the national life of the country was an oft-repeated theme. Their imperfect integration, and the reasons for it, became so widespread a topic for reflection and consternation that it came to define an elemental aspect of modern German culture. It is little wonder, then, that the physical, psychological, and cultural characteristics of Jews became omnipresent themes in German anthropology, because it was felt that perhaps the "science of man" could provide an answer to the troubling persistence of the Jewish community long after their emancipation and acculturation. Conversely, the absence of the Jewish theme from British anthropology is one indicator of the acceptance of Jews in English society.

Another reason for the absence of Jews from the early literature of British anthropology has to do with nineteenth-century ideas of sociocultural evolutionism and, by extension, their relation to biological or "racial" status. In turn, these ideas must be considered within the context of contemporary discussions of progress and civilization. In nineteenth-century Britain, this debate was shaped by such facts of British life as industrialization, laissez-faire economics, parliamentary democracy, and a middle-class Protestant culture of self-discipline and strict morality.⁴ By the mid-nineteenth century, it was to these very institutions (the defining forces of their civilization) that British social thinkers referred when attempting to discern the differences between the English and what were at the time called the primitive peoples. By intense anthropological scrutiny of these "backward" Others, who were bereft of the fundamentals of (Western) civilization, the British were able to examine, identify, and define themselves. Since Jews took part in all these institutions of British life (with the exception, of course, of Protestant Christianity), they were not perceived as a possible threat to the system and the cause of national identity building. Even after thousands of Russian Jews began pouring into the East End of London in 1881, thereby threatening, in the minds of anti-immigrationists, these hallowed national institutions, the appearance of Jews in British anthropological literature was short-lived and of minor significance.

To appreciate how the picture of the Jews evolved, or more accurately, failed to evolve in British anthropology, it is crucial to examine this literature in the light of the social circumstances in which Jews found themselves upon their readmission to England in the seventeenth century. At that time their role as loyal, law-abiding citizens, serving England in its ongoing commercial and political rivalry with Spain and Holland, ensured that future generations of English Jews would be treated not as trespassers on the national landscape but as legitimate and welcome heirs of British culture.[5] Thus, it is not surprising that British anthropology, from the late eighteenth century to the middle of the nineteenth, which tended to focus on more "exotic" peoples, should ignore a group so settled, so acculturated, so English, as the Jews. This is not to say that the Jews were completely ignored in early British travel or ethnographic literature. The issues of the Jewishness, origin, and descent of *Homo judaeus* were most certainly addressed, but never examined in any systematic or scientific way. In fact, until the middle of the nineteenth century, descriptive or proto-anthropological discussions of Jews and things Jewish were generally provided by those who were heavily involved in millenarian or conversionist activity.

Two of the earliest writers on a topic of Jewish anthropological interest were Jews. In 1641, the Marrano adventurer Antonio de Montezinos claimed to have discovered near Quito, Ecuador, a tribe of Indians belonging to the Hebrew tribes of Reuben and Levi and practicing various Jewish rites and customs. Upon his return to Holland, Montezinos rendered an account of his story to Rabbi Menasseh ben Israel, the distinguished scholar of the Dutch Jewish community and ambassador of European Jewry to Christian Europe. Menasseh accepted the report of Montezinos, and in *The Hope of Israel* (1650) he not only repeated the claims of the intrepid explorer but laid out the broader, millenarian implications of the startling discovery. The author invoked the authority of the prophet Daniel, who held that the final redemption would commence only when the total dispersion of the Jewish people was complete. Now that descendants of the Lost Tribes of Israel had been located in the Americas, it only remained for the Jews to be reintroduced into England. Menasseh ben Israel dedicated his book to the British parliament and submitted it along with his petition that the Jews be readmitted to England. With this achieved, the millennium would be at hand. This theory gained even wider currency with the publication, also in 1650, of Thomas Thorowgood's *Iewes in America*, which brought before the English reading public Montezinos's testimony of American Jewish tribes. Thorowgood believed that the discovery of Hebrews in America was but a prelude to first Indian and then Jewish conversion to Christianity.[6]

For the British, the connection between native Americans and Jews did not fade as the tide of millenarianism receded. As late as 1775 the issue was given a more thorough and "scientific" treatment with the publication of James Adair's classic, *The History of the American Indians*. Long regarded as a fundamental text for historians and ethnologists, Adair's *History* is justly famous for its long exposition of "Observations, and Arguments"—twenty-three—"in Proof of the American Indians' Being Descended from the Jews." Employing a comparative approach, the author detected, amongst other things, linguistic, religious, military, medical, judicial, calendrical, and even sartorial similarities between Jews and Indians.[7]

Adair's was merely one of the most illustrious and scholarly treatments of this widely held anthropological theory whose lineage can be traced back to Gregorio Garcia's *Origen de los Indios* in 1607. Other fervent adherents of the idea included Cotton Mather, Roger Williams, and William Penn. Perhaps the theory's last and most elaborate treatment was in Viscount Edward King Kingsborough's *Antiquities of Mexico,* which appeared in nine volumes between 1830 and 1848. The issue of descent and racial ancestry not only hinged on wild theories that united the peoples of the New and biblical worlds, nor did it necessarily have millenarian implications. As with other modern theories of ethnic descent and collective identity, this claim was a careful blend of ancient pagan beliefs and Christian doctrine invented to help explain or justify a contemporary political reality or glorify a particular group.

Most European nations have created such heroic myths of origin. From Spain to Russia, the peoples of Europe sought a pedigree that would bolster chauvinistic views of what they saw as their own God-given superiority. An ancient and noble lineage, liberally peppered with heroes and maidens of valor, are necessary ingredients in the preparation of a great national history or myth of national origin.

Christianity helped perpetuate these European myths of origin by stressing that the common ancestor of all mankind was Adam. The line of descent was direct to Noah but then branched at Noah's three sons, Japheth, Shem, and Ham, and from that point became blurred. This genealogy was in turn taken up by the Church Fathers, who mixed them with local pre-Christian and regional folkloric tales of origin. The result was that Europeans were free to embellish their genealogies to suit their own personal and national agendas. Thus, people could detect their national origins in the Goths, Celts, Franks, Gauls, Saxons, Teutons, Aryans, and Slavs, and revel in the supposed racial attributes of their tribal ancestors.[8] Only the English, though, chose to retain the myth of their Hebrew origins alongside their other celebrated Greco-Roman, Celtic, and Germanic ones. This special place for Jews in British racial folklore clearly contributed to the unfolding of a harmonious relationship between the English and the Jews.[9]

John Toland was one person whose tone and attitude helped define that relationship. A man of the Enlightenment, Toland championed the cause of naturalizing the Jews by appealing to human reason and British self-interest. In speaking of the benefits to the nation if naturalization were granted, Toland also stressed the commonality of descent and outlook shared by Englishmen and Jews. In his *Reasons for Naturalizing the Jews in Great Britain and Ireland* (1714) he argued that because ancient Jews had been "Shepherds in Mesopotamia, Builders in Egypt, and husbandmen in their own Country" as well as "anciently excellent Soldiers," and because "their maritime Tribes produc'd as good Seamen" as the English, they would bring these skills with them to the British Isles.[10] Toland suggested that the ancient characteristics of Jewish national organization lay dormant in Jews and could be made to appear only in a republic of reason and tolerance. This is an example of the inherent ambivalency of Enlightenment thinking on issues concerning national progress, human difference, and individual capacity. Toland's reference to inherent Jewish characteristics highlights this tension.

Yet Toland was no innatist. He rejected the notion that genius or any other "bent of mind" was the property of any one "Family or Nation." If either exists, "it wholly proceeds from Accident, and not from Nature." He went on to claim that Jews did not differ in character from the other peoples among whom they lived. The most significant causes of the differences of human nature were not heredity or climate, as some had suggested, but rather "the different methods of Government and Education." In the Diaspora, Jews had no "common or peculiar inclination" distinguishing them from other peoples "but visibly partake of the Nature of those nations among which they live, and where they were bred." Toland could be a paragon of rationality and fair-mindedness holding out against the forces of prejudice: "I know a person, no fool in other instances, who labor'd to perswade me, contrary to the evidence of his own and my eyes (to mine I am sure) that every Jew in the world had one eye remarkably less than the other, which silly notion he took from the Mob. Others will gravely tell you, that they may be distinguish'd by a peculiar sort of smell, that they have a mark of blood upon one shoulder, and that they cannot spit to any distance, with a world of such extravagant fancies, exciting at once laughter, scorn, and pity."[11]

Toland himself, however, was not totally immune from silly notions. He suggested that another sound reason for naturalization was that the English owed their physical descent to the Jews. Observable features of their Hebraic pedigree could, argued the deist Toland, be observed even today: "You know how considerable part of the British inhabitants are the undoubted offspring of the Jews and how many worthy prelates of this same stock, not to speak of Lords and commonors, may at this time make an illustrious figure among us. . . . A great number of 'em fled to Scotland which is the reason so many in that part of the Island have such a remarkable aversion to pork and black puddings to this day, not to insist on some other resemblances easily observable."[12] Indeed, belief in the descent of the English from the Jews gained such wide currency in the nineteenth century that a movement called the British-Israelites claimed hundreds of thousands of followers, even counting Queen Victoria and King Edward VII among its patrons.[13] The group relied on dubious interpretations of the Scriptures and even fanciful etymological tricks to make its case. The most creative was the origin of the word *British* itself, which they said was derived from the Hebrew *b'rit ish*, "covenant of mankind." But despite honoring Jews interminably, the movement was somewhat antisemitic. Although its members trumpeted their supposed ancestral link with ancient Israelites, they drew a sharp distinction between themselves and modern Jews, with whom they they felt no ties.

Well into the nineteenth century, English missionaries continued to provide ethnological accounts of Jews they encountered in their travels. These reports were generally unsystematic and devoid of scientific methodology. Nonetheless, the more sophisticated of them, such as the Reverend Claudius Buchanan's *Christian Researches in Asia* (1811), allow a glimpse of developments to come in British ethnology. Buchanan drew upon a host of methods that enabled him to make racial comparisons and answer questions concerning racial origin, descent, and heredity. He used comparative philology and linguistics, history, and biblical analysis to study human difference. This interdisciplinary approach to race science remained in vogue in England until the middle of the nineteenth century, when it was replaced with a more purely physical analysis of racial types.

Buchanan was vice-provost of the College of Fort William in Bengal. He claimed that in the course of his Indian travels on missionary work he had heard of "distinct colonies" of Jews living as "a distinct and separate people" among the Hindus, under constant persecution yet never destroyed, "burning, like the bush of Moses, and not consumed." And so from November 1806 to February 1807, and again in January 1808, Buchanan set out "to hear the sentiments of the Jews from their own lips." What follows is a fascinating account of a little-known Jewish community near Cochin, in the Malabar district. Buchanan discussed the origins of Indian Jewry and its arrival soon after the destruction of the second Temple in 70 CE. He stated that he had personally examined "their ancient brass Plate, containing their charter and freedom of residence, given by the king of Malabar . . . [who] . . . granted them a place to dwell in called Craganor. He allowed them a patriarchal jurisdiction within the district, with certain privileges of the nobility."

Buchanan drew a sharp distinction between what he called the "White, or Jerusalem Jews" and the "Black Jews" whose appearance indicated to him that they had broken away from the "parent stock in Judea" and had intermarried with non-Jews. So thoroughly assimilated had these Jews become that Buchanan found them to be indistinguishable from the Hindus. The physical and cultural affinity of the Black Jews with Indians was such that even "the White Jews look upon the Black Jews as an inferior race, and not of *pure* cast."[14]

Buchanan's contribution to modern anthropology is substantial. The racial duality that he observed among Jews made him one of the very first in modern British anthropology to raise the issue of there being more than one "Jewish type.[15] This theory became widely accepted by the middle of the nineteenth century. Buchanan's assertion that thus far the environment had had no effect on the appearance of the "White Jews" was used by later adherents of polygenism. Since Jews were believed to have remained pure, the existence of Jews with any complexion that deviated from the

"Jewish norm" meant that this anomaly must be derived from another racial source.

Buchanan is also important because of the later misappropriation of his findings. Misuse of statistics or the subtle manipulation of evidence, either by accident or by design, was a staple of the entire enterprise of race science.[16] Buchanan, for example, was wrongly interpreted by the champions of polygenism, who used his findings to dispute the unity of origin of these so-called White and Black Jews. The authors of one of the most famous books on race in the nineteenth century, the Americans Josiah C. Nott and George R. Gliddon, argued that Buchanan's evidence was proof that the Black Jews had been "adulterated by dark Hindoos—Jews in doctrine, but not in stock."[17] But it is clear from Buchanan's testimony that he had only suggested that intermixture between Jews and Hindus had taken place in India. Nowhere did he imply that both groups of Jews were not originally from the same stock. Buchanan was simply unable to accept the validity of any theory of racial difference that challenged the authority of the Bible regarding humanity's unity of descent. The attempt to claim Buchanan as a polygenist entailed a deliberate distortion of his beliefs.

Early British ethnology was guided by what Stocking has described as "religious commitment and underlying biblical orthodoxy."[18] The work of men like the missionary Buchanan, fundamentally religious and yet groping for scientific method, helped pave the way for, and influenced, the first generation of British anthropologists, such as James Cowles Prichard.[19] Prichard's innovative method of analyzing human difference by means of botany, comparative ethnology, and comparative philology dominated British racial science for at least the first half of the nineteenth century. In his seminal book of 1808, *Researches into the Physical History of Man*, Prichard drew on a vast array of works from the classical world, eighteenth-century exploration literature, new writings on the biological sciences, comparative anatomy, and the philological discoveries of the great Orientalist William Jones and his colleagues at the Asiatick Society.[20] Yet despite its standing as the premier "modern" British anthropological text, one searches in vain for any extended discussion of the Jews. When Prichard did address the anthropology of the Jews, it was in a most limited and specific way—once again confirming the claim that Jews as a people were of little general interest to British anthropologists.

Prichard first mentions Jews in the fifth chapter of his *Researches*, "On the Causes which have produced the diversities of the Human Species," but they are treated of within the context of the debate over the possible correlation between climate and skin color. The standard view was expressed by the French naturalist and superintendent of the Jardin du Roi, Georges Louis Leclerc de Buffon: "Where the heat is excessive as in Senegal and Guinea the men are perfectly black; where it becomes somewhat temperate, as in Barbary, Mogul, Arabia &c. the men are only brown; and lastly, where it is altogether temperate as in Europe and Asia the men are white."[21] Jews became a litmus test of this theory well into the twentieth century. Nearly all who believed that climate affected complexion saw in the Jews living proof of their theories, for the Jews were regarded as a perfect example of a pure race, widely dispersed and yet physically resembling the peoples amongst whom they live. Since no intermingling of any kind had taken place, what else but the influence of climate could account for the variations in their skin color and their physical similarity to the natives?

Prichard dismissed the idea that Jews are a unique example of a dermatologically adaptive people. In this he took issue with the ex-president of the College of New Jersey (Princeton), Samuel Stanhope Smith, whose *Essay on the Causes of the Variety of Complexion and Figure in the Human Species* (1787) had become a proof text for anthropologists who held that climate alters complexion. According to Smith, "no example can carry with it greater authority on this subject than that of the Jews. Descended from one stock, prohibited by their most sacred institutions from inter-marrying with strangers, and yet widely dispersed into every region on the globe, this one people is marked with the peculiar characteristics of every climate. In Britain and Germany they are fair, brown in France and in Turkey, swarthy in Portugal and Spain, olive in Syria and Chaldea, tawny or copper-coloured in Arabia and Egypt."[22] Prichard directly disputed Smith's position as "an inaccurate statement of the facts." Intermarriage between Jews and other peoples had indeed taken place, and it accounted for variations in skin color; where Jews had not intermixed, they had retained their "primitive complexion." Prichard tries to reconstruct this original hue on the basis of passages from the Scriptures that, in his view, suggest that "the Jews in the time of their monarchs of the house of David resembled the inhabitants of the South of Europe in their complexion. They had bushy black hair and a white skin, with some variety probably as we see in all races, and acquiring a darker hue in consequence of exposure to heat and air" (182–183).

Prichard challenged the view, derived from far too literal a view of Jewish religious legislation and social custom, that all Jews had remained physically isolated from their neighbors. In doing so, however, he opted for an argument that stressed the fixity or unchanging uniformity of the Jewish type, particularly as it pertained to color. And it was this very point that led him to differ so sharply from Smith. Prichard claimed that the nations among whom the Jews originated were white and therefore so too were the Jews, because there was no way they could take on a different hue from their non-Jewish neighbors. Both were equally unsusceptible to the effects of climate.

To prove the limited effect of climate on complexion, Prichard offered the example of English Jews, who, if the climate theory were true, should have become fairer in the gray and temperate climate of Britain. But such was not the case: the Jews of England had lost neither their black hair nor the "choleric and melancholic temperaments" that so distinguished them from the "sanguine" English (186). Prichard put forward one more piece of evidence to bolster his claim: Buchanan's case of the black and white Jews of Malabar. In accepting this particular testimony of Buchanan, Prichard seemed to arrive at a biological definition of the Jews. The "real" or "true" Jews were white, and he claimed that "the white complexion will be permanent during any length of time." (187)

By maintaining that the Jews retained their original coloring, Prichard rejected the theory of the inheritance of acquired characteristics. The effects of climate, he declared, cannot be passed on from generation to generation. In this context, Prichard also raised the issue of circumcision, because in nineteenth-century Europe it marked the Jewish body as fundamentally different from that of the non-Jew. To medical science, the circumcised Jewish male became the exemplary Jew: in the scientific literature, when physicians and race scientists spoke of Jews they usually meant Jewish males.[23] It was on account of the centrality of this ritual and the role it played in the creation of European images of the Jew that Prichard relied on it when arguing for the limited influence of environment on humans. He asserted that the practice of circumcision by Jews and other nations for thousands of years had not yet led "the artificial state [to] become natural" (199).

It is difficult to know exactly what Prichard's intention was in denying the possibility of there being "true" Jews who were black. But a look at his religious beliefs and his attitude toward blacks may shed light on the matter. Prichard was a Quaker turned Evangelical, and his scientific work was informed by his religious outlook and his strict adherence to biblical orthodoxy. This may have included a strain of British-Israelism. Thoroughly learned in the Bible, and with a sound knowledge of Hebrew, Prichard very likely saw himself as a spiritual descendant of ancient Israel.

At the same time, although he was a committed monogenist till the end of his life and a firm supporter of the Aborigines Protection Society when it was established in 1837, Prichard saw blacks as morally and intellectually inferior. In his early works he argued for the original blackness of the human race and was of the opinion that humans had "progressed" from barbarism to civilization, and similarly from blackness to whiteness. Despite this, Prichard was not a scientific racist, and eventually he became a target of men who were; nevertheless, he spoke of blacks with the contemptuous tone that was typical for his day. Thus, his philosemitism coupled with his arrogance toward blacks may have unconsciously led him to insist on the "true" and "original" whiteness of Jews, a theory that accorded well with his religious inclinations, moral rectitude, and aesthetic sensibilities.

The relegation of the Jews by Prichard to a circumscribed role in the climate-complexion debate was continued in the work of Sir William Lawrence.[24] In 1819 Lawrence, a London surgeon, published his justly famous *Lectures on Physiology, Zoology, and the Natural History of Man*. Monogenist in conclusion and interdisciplinary in methodology, the work of Lawrence stands as a companion to that of Prichard and has secured for him a place among the founders of British anthropology. Lawrence's and Prichard's works amount to encyclopedias of the races of most of the world known to them, yet just as Prichard had done, Lawrence limited discussion of the Jews to the debate over climate and complexion.

Jews first appear in Lawrence's *Lectures* in chapter 9, which bore a similar title to Prichard's, "On the Causes of the Varieties of the Human Species." Rejecting the climate theory by dismissing the paradigm of the inheritance of acquired characteristics, Lawrence also adduced as proof of the untenability of the hypothesis the example of circumcision. "After the operation of circumcision has prevailed for three or four thousand years, the Jews are still born with prepuces, and still obliged to submit to a painful rite."[25] This singular example would be repeated by many anthropologists into the twentieth century.

Lawrence described the Jews as a race that displayed an unusual stability of type and had remained totally unaffected by the climates in which they dwelled. In fact, despite their dispersion, they "exhibit one of the most striking examples of national formation [because] their peculiar religious opinions and practices have kept the race uncommonly pure." Lawrence maintained that those anthropologists who had used the Jews as an example of a race whose appearance was modified by the diverse climates in which they lived had erred. And to underscore the point, he, like Buchanan and Prichard, asserted that the Jews "have naturally a white skin and the other attributes of the Caucasian race. In hot countries they become brown by exposure, as an European does, but they experience no other influence from climate."[26]

In sum, early British anthropology's two leading lights recognized in the Jews three major characteristics: their social separateness, their supposed racial purity (a claim that they never attempted to prove), and their geographic dispersion and paradoxical similarity to surrounding populations. These characteristics were offered as evidence of the fallacy of the climate-complexion theory. Generally, however, they found nothing particularly remarkable nor interesting about the Jews from a broader anthropological point of view. The passing and exemplary discussions of Jews in the scientific texts of Prichard and Lawrence mirrored in some respects the unobtrusive character of Anglo-Jewry as a whole and the somewhat lackadaisical English attitude towards the country's Jewish subjects.

By the middle of the nineteenth century the structure of British anthropology began to change, and consequently so too did the way the discipline addressed the Jews. Until this time, British anthropology was characterized by the humanitarian-religious-philanthropic triad, with roots at least extending to the seventeenth century and the Protestant Cromwellian Protectorate. This older variety, which was informed by biblical orthodoxy, combined physical anthropology with comparative linguistic and historical analysis. It was superseded by a new kind that was more stridently biologically determinist, uncompromising, and pessimistic.

Two features distinguish the new anthropology from the old. First, it essentially dispensed with that which was not strictly 'scientific,' that is, philology and history, because they lacked an empirical base. Their place was taken by statistics and comparative anatomy. The result was an anthropology that was increasingly innatist and ever more racist. Second, the new anthropology was spurred on by heightened interest in the "races" of Europe. The tumultuous revolutions on the Continent in 1848, which also reverberated in England, gave rise to a kind of "us" versus "them" thinking in which European class struggles were identified by some thinkers—Gobineau is the most famous—as actually racial ones.[27]

The changing over of the old anthropological guard to the new is best illustrated in two works that were both published in 1850: Robert Gordon Latham's *The Natural History of the Varieties of Man* and Robert Knox's *The Races of Men*. Following Prichard's death in 1848, the mantle of leadership of the Ethnological Society of London passed to Latham.[28] Like Prichard, he was a physician who adopted a multidisciplinary approach to his anthropological work by remaining true to the philological-historical method. According to Latham, each of the three widely accepted types of human being—Mongolian, African, and European—had a series of subdivisions. He classified the Jews in the African division, which he called Semitic Atlantidae, along with Syrians, Assyrians, Babylonians, Beni Terah, Edomites, Beni Israel, Samaritans, Arabs, Ethiopians, Canaanites, and Malagasi. It is not surprising that all these peoples (some of whom were white) were classed as Africans, for as Latham said, "No fact is more necessary to be remembered than the difference between the Negro and the African."[29] He maintained that while it was not possible to claim that a dark complexion was an exception rather than a rule in Africa, the hue of the Arab, Indian and Australian is the prevalent color.

In his work, Latham concentrated heavily on comparative linguistics and history but also gave a more detailed description of the Jews than any of his English predecessors. In fact, he said nothing of their value in refuting the climate-complexion theory and focused more on their physical and moral features. Jews, he wrote, differ in appearance from the Arabs in that they possess a "greater massiveness of frame" and "thicker lips," while their "nose[s] are more frequently aquiline" and the "cranium [is] of greater capacity."

Latham described two other aspects of Jewish civilization: intellectual culture and moral influence. Of the former, he noted that Jews developed "preeminently early" and had remained an intellectual culture throughout their history, even up to the present, after their adoption of secular culture and vernacular languages. By moral influence Latham was referring to the religious system created by the Jews and its impact on world history. In giving birth to Christianity and Islam, Judaism passed on its moral and ethical foundations to other nations.[30]

The adoption of this dualistic approach, which both described the Jews and speculated on their place in world history, puts Latham in a kind of intellectual limbo. On the one hand, the roots of his anthropology are in Prichard and Lawrence. On the other, his physical descriptions coupled with his analysis of the historical role of the Jews (and the other races he examined) places him in the company of the mid-century theorists Comte Arthur de Gobineau in France and Robert Knox in England,[31] who championed a theory of history which held that race, and not environment, was the causal factor in human affairs.

Latham was not a racist like both Gobineau and Knox, but his work showed more sophistication than his mentor Prichard's by relying far less heavily on the literalness of Scripture and the inviolability of traditional religious thought. In the end, however, the Jews were of scant importance to Latham. He spent very little time talking about them because his emphasis was still on the "exotic" races. Nevertheless, his dualistic approach to the Jews as a race, his physical descriptions of them, and his analysis of their historical value makes Latham a transitional figure serving as a bridge over the increasingly turbulent waters of British physical anthropology.

With Robert Knox, the bridge was crossed from Prichardian anthropology to biological racism. The appearance of his *Races of Men* (1850) made Knox the preeminent representative of the "new" anthropology. This work, unlike Gobineau's, was of immediate influence and was crucial in staking a claim for the growing anti-environmentalist, polygenist school of anthropology.[32] Knox had been a celebrated anatomist in Edinburgh, one of the premier European cities for medical studies; his pupils had designated him "Knox primus et incomparabilis."[33] His professional star burned out, however, in 1829, when he became embroiled in a scandal concerning the murderous activities of William Burke and William Hare, who were accused of killing innocents and selling the cadavers for use in anatomical studies.[34] Leaving Scotland in disgrace after being named as a purchaser of the bodies, Knox finally settled in London, where he earned a living lecturing and translating foreign medical texts.

Races of Men appeared in the wake of the mid-century revolutions. On the Continent, order and industriousness had replaced chaos and heady aspirations. Even in England, where revolution did not erupt, the new spirit of realism and materialism was keen. In 1850 John Paxton completed his monument to British ingenuity, industry, and predominance, the Crystal Palace, and in the same year Herbert Spencer's *Social Statics* outlined an evolutionary concept of society that explained human progress as the outcome of struggle and survival.[35]

Disillusioned by the events of 1848, Knox, like many writers and thinkers of the time, sought reasons for the great European upheavals. For this embittered Scot, the key to understanding recent events was race; indeed, the aim of his book was to show that "in human history race is everything." Impelled to prove that wars and revolutions were the products of racial and not class conflict, Knox was the first European to write a biology of cultural despair. In contrast to the optimist Spencer, he derided all previous theories of progress and mocked contemporary notions of civilization: "Look at Europe; at either bank of the Danube; at Northern

47

Africa; at Egypt; at the shores of the Mediterranean, generally, and say what progress civilization has made in these countries since the decline of the Roman Empire."[36] He even went as far as to challenge the fundamental assumption that Christianity was responsible for the march of progress in Europe. That which others took as a sign of material progress, "empires, monarchies, nations," was for Knox "human contrivances often held together by fraud and violence" (11–12). This overall view of human history, in which he saw nations and empires overextending the limits of their racial capacities, had direct bearing on his opinions of Jews.

Knox was the first British anthropologist to pay Jews anything more than passing and limited attention. He spent a summer on the streets of London and Amsterdam observing Jews, mainly in the poorer quarters, and the vividness of ghetto life left an indelible impression on his anthropology. But that impression is not positive: Knox's antisemitic vitriol comes through clearly in his writings. Rather than commence *Races of Men* by giving a purely physical description of his subjects—a common scientific procedure—he immediately launched into a rambling, familiar, and wholly unscientific tirade against the Jewish character and the limited scope of the Jews' occupational structure. "But where," he asked, "are the Jewish farmers, Jewish mechanics, labourers? Can he not till the earth, or settle anywhere? Why does he dislike handicraft labour? Has he no ingenuity, no inventive power, no mechanical or scientific turn of mind?" (131). How dissimilar these Jews were from Knox's superior race, the Saxons, who were "thoughtful, plodding, industrious beyond all other races, a lover of labour for labour's sake" (44–45). Given all their deficiencies, the Jews would be incapable of building their own Crystal Palace.

Knox was also responding to the greatest European champion of the concept of the Jewish race: Benjamin Disraeli. In the section of his book where he lists the deficiencies of the Jewish race, Knox took aim at one of Disraeli's novels, *Coningsby,* whose theory of Jewish superiority he derided as "not merely a fable as applied to the real and undoubted Jew, but . . . absolutely refuted by all history" (132). In this first volume of the Young England trilogy, Disraeli's unbridled, chauvinistic pride in his racial patrimony shines forth as Sidonia, the Jewish banker, explains the meaning and significance of Jewishness to the naive, Cambridge-bound Coningsby. Sidonia notes the irony of his being called upon to bail England, which does not grant him full citizenship, out of its financial difficulties. But in England, he says, these disabilities are minor, little more than an annoying humiliation insufficient to break the back of a great race: "Do you think that the quiet humdrum persecution of a decorous representative of an English university can crush those who have successively baffled the Pharaohs, Nebuchadnezzar, Rome, and the Feudal ages? The fact is, you cannot destroy a pure race of the Caucasian organisation. It is a physiological fact; a simple law of nature. . . ."[37]

Disraeli reserved his greatest bombast for Sidonia's discourse on the musical abilities of the Jews. After affirming the omnipresence of Jews in all the affairs of Europe, religious, political and cultural, Sidonia soliloquizes at great length on the musical preeminence of the Jews: "The ear, the voice, the fancy teeming with combinations, the imagination fervent with picture and emotion, that came from the Caucuses, and which we have preserved unpolluted, have endowed us with the almost exclusive privilege of MUSIC; that science of harmonious sounds, which the ancients recognised as most divine, and deified in the person of their most beautiful creation." The choruses and orchestras of Europe are filled with Jews who have hidden their Jewish names to "conciliate the dark aversion" of the gentile peoples among whom they live and work; when the audience applauds an opera by Rossini or Meyerbeer, or "thrill into raptures at the notes of an aria by a Pasta or a Grisi, little do they suspect that they are offering their homage to 'the sweet singers of Israel!'" Indeed, declares, Sidonia, "musical Europe is ours."[38] This is Disraeli at his most chauvinistically rhapsodic, and is a classic example of the nineteenth century treatment of the race concept in literature.[39] It also refutes the charges of tone-deafness and artistic plagiarism as inherited Jewish defects that had become staples of modern antisemitic propaganda.[40]

Writing in 1844, Disraeli was merely reacting to the antisemitic rumblings that were already coming from the artistic world. But soon these rumblings turned into an explosion, and supposed Jewish inferiority became a frequent topic in literature and music. In 1850 Richard Wagner published the viciously antisemitic and influential *Das Judenthum in der Musik*.[41] Inspired by his professional rivalry with the Jewish composers Giacomo Meyerbeer and Felix Mendelssohn, and his belief that German music's deterioration was accompanied by the rise of Jewish predominance in the field, Wagner constructed an entire racial fantasy that dismissed all Jews as bereft of musical abilities. Furthermore, their racial difference prevented them from partaking in and contributing to European culture. They were unable to compose music because they lacked passion and were singularly devoted to the lure of money. In short, their alienness from the *Volk* community prevented them from producing truly original works of art: "Our whole European art and civilization . . . have remained to the Jew a foreign tongue."[42]

Wagner's work did not go unechoed abroad. Knox also dismissed the artistic abilities of the Jews, claiming "that the real Jew has no ear for music as a race, no love of science or literature." (131) He scoffed at Disraeli's catalog of Jewish overachievers and refused, out of "respect for scientific truth," to stoop to refuting "the romances of Disraeli." But having examined the list anyway, he declared that he had not "met with a single Jewish trait in their countenance . . . and, therefore, they are not Jews, nor of Jewish origin" (140). Thus, it is in the context of Disraeli's mindless boasting, and the hate-filled declamations from Bayreuth, that we must read Knox's attacks on the aesthetic deficiencies of the Jews.

The real importance of the sections pertaining to the Jews in *Races of Men* lies not with Knox's banal appraisal of Jewish cultural contributions. Rather, *Races* contains British race science's first sustained and comprehensive attempt at describing the physical appearance of the Jews. The Jew presents "three different casts of features: the first Jewish, par excellence, and never to be mistaken; a second, such as Rembrandt drew; and a third, possibly darker, of other races intermingled" (133). Indeed Knox discovered that all races present three different forms and therefore in this respect the Jews were not unique.

In his description of the first type, or "form," of Jew, Knox advanced a position on Jewish appearance that was repeated with great tenacity down to the middle of the twentieth century: viewing the Jew as black. The contour of the Jew's face, he wrote, "is convex; the eyes long and fine, the outer angles running towards the temples; the brow and nose apt to form a single convex line; the nose comparatively narrow at the base, the eyes consequently approaching each other; lips very full, mouth projecting, chin small, and the whole physiognomy, when swarthy, as it often is, has an African look" (133). If Jews shared many physical similarities with blacks, then they could be safely segregated in the mind's eye from the majority population. In the modern period there was a practical necessity for doing this. In countries where assimilation proceeded at a steady pace, and where indigenous Jews were indistinguishable in appearance from the majority population (for example, England and Holland, where Knox was conducting his fieldwork), highly artificial and fanciful distinctions were made by anthropologists.[43] Jews, who had always stood apart by virtue of religion, language, dress, occupation, and residence restrictions, were diminishing in number in Western Europe. Most Jews, having successfully blended in, were by their very sameness a threat to the idealized racial integrity of the state. By giving the Jew black physical features, men such as Knox were able to distinguish and separate Jews from white Europeans just as obviously as black Africans differed from white Europeans.

There is little to indicate whether Knox found the Jews' "African look" particularly attractive or repulsive. His description is a fairly standard, dispassionate, and for its time rather scientific evaluation. However, Knox soon qualified it. The features that Knox termed "fine" appear only in the young; with age, they become exaggerated. As the Jew ages, he metamorphoses into an ugly and repulsive creature easily identified in the crowd of Saxons, who are "a tall, powerful athletic race of men; the strongest as a race on the face of the earth, [having] fair hair, with blue eyes, and so fine a complexion, that they may almost be considered the only fair race on the face of the globe" (43). The Jew descends rapidly on the scale of human beauty as the "energies of the chest and the abdomen, . . . the stomach and the reproductive systems, . . . the overdevelopment of the nose and mouth [dislocate] by their larger development that admirable balancement of head and face, of brow and nose, eyes and mouth, cheeks and chin." Balance and proportion, order and uniformity, are hallmarks of classical beauty; in the mature Jew these features are absent. Likewise, the once stable European political and social order had, since the revolutions of 1848, lost its own "admirable balancement" and become chaotic and distorted. To the despairing scientist of the mid-nineteenth century, the decline of a race became a metaphor for the decline of a world.

After praising the transient beauty of the youthful Jew, beauty that exists "provided your view is not prolonged," Knox draws a portrait of the middle-aged, assimilated Jew. Having initially described his want of character and aesthetic sensibility, Knox now deprives him entirely of any beauty and, indeed, any Europeanness. The Jew is now an alien, an object of horror and prognathous repulsiveness:

> Brow marked with furrows or prominent points of bone, or with both; high cheek-bones; a sloping and disproportioned chin; an elongated, projecting mouth, which at the angles threatens every moment to reach the temples; a large, massive, club-shaped, hooked nose, three or four times larger than suits the face—these are features which stamp the African character of the Jew, his muzzle-shaped mouth and face removing him from certain other races, and bringing out strongly with age the two grand deformative qualities—disproportion, and a display of the anatomy. Thus it is that the Jewish face never can, and never is, perfectly beautiful. (134)

The theme of the Jew as black was most widely subscribed to in Germany from the late nineteenth century through the Third Reich. One of the most infamous examples of this type of thinking was when Karl Marx called his rival, the Jewish socialist leader Ferdinand Lassalle, "the Jewish nigger." On July 30, 1862, Marx wrote to Engels of Lassalle: "It is now fully clear to me—as the shape of his head and the way his hair grows proves—he is descended from the negroes who joined Moses in his flight from Egypt (unless his mother or paternal grandmother were crossed with a nigger). Now this union of Jew and German with basic negro stock must produce a strange product. The fellow's importunity is also nigger-like." The same idea was expressed by such disparate figures as the racist thinker Houston Stewart Chamberlain, who claimed that the Jews of Alexandria had undergone an admixture of "Negro blood," and Otto von Bismark's friend, the Prussian state official Hermann Wagener, who referred to Jews as "white negroes." He naturally lamented, however, that they lacked the "robust nature and the capacity for physical work of the Negro."[44]

In denying the assimilability of the Jew, Knox stands alone among British anthropologists, operating more within the traditions of Continental race science and antisemitism. In Germany, it was the assimilated Jew who became the target of hostile forces. From the 1780s on, opponents of Jewish emancipation seriously argued that Jews, no matter how integrated they had become, were unable to divest themselves of their indelible Jewish (especially mental) characteristics and become Germans.[45] This was not the case in England. Moreover, such antisemitic notions as did exist never translated into the institutionalization of prejudice, as was the case in Central Europe. By the 1850s in England, only the requirement of taking a Christian oath to hold public office remained as the final hurdle before full emancipation. After a compromise agreement was reached permitting each house to administer its own oath, Baron Lionel de Rothschild was seated in Parliament in 1858.

Knox's veiled reference to Disraeli and other assimilated Jews—"losing sight of his origin for a moment, he dresses himself up as the flash man about town; but never to be mistaken for a moment—never to be confounded with any other race"—is but an echo of Continental concerns and is out of touch with prevailing views in Great Britain. The contrasting attitudes of the English and the Germans towards the Jews was noted by contemporaries. In his address to the Anglo-Jewish Historical Exhibition of 1887, the Jewish historian Heinrich Graetz summed up the differences between Jewish emancipation in England and Germany and, by extension, the very nature of what it meant to be Jewish in those countries: "You [English Jews] had not to give solemn assurance that you are good patriots, that you love your native country as much as your fellow citizens. . . . The honours you have received have been granted to you sans phrase as the descendants of Jacob, as the guardians of your ancient birthright."[46]

Although Knox is intellectually closer to his European counterparts than to his British ones, unlike them he was not a national chauvinist. He never questioned that Jews or another people could become English. This is not to deny that Knox was a racist. He dismissed terms such as *English, Teuton, German, French*, and *Russian*, preferring *Saxon, Celt, Sarmatian*. To Knox, the difference was vast. Nationalities, empires, and monarchies are plastic and ephemeral, "human contrivances often held together by fraud and violence." Races, however, are stable and permanent. Any people can become members of a nation and display its national characteristics, but they can never, through the artificial process of assimilation, become members of a race into which they were not born.

Anthropologically, Knox asserted, race was everything. "If anyone insists with me that a Negro or Tasmanian accidentally born in England becomes thereby an Englishman, I yield the point; but should he further insist, that he, the said Negro or Tasmanian, may become also a Saxon or Scandinavian, I must contend against so ludicrous an error" (12). Similarly, be they Jewish men who "saunter about Cornhill in quest of business" or Jewish women who are "the beauties of Holywell-street," they may succeed in becoming English but will never become part of the Saxon race. This type of thinking was rare among German and French race scientists who hardly distinguished between race and nationality.

This distinction has direct bearing on Knox's attitudes toward Jews. To Knox, races were fixed and permanent, ideally adapted to their original environments. Races could not uproot themselves, move to another part of the globe that had a completely different climate and geography, and expect to survive. For this reason, Knox was an outspoken critic of colonization, hurling vehement invective at contemporary empires and reserving scorn and contempt for those of the past. According to him, the territorial lust of the ancients ended in failure, and the same acquisitive nature that drives the moderns is also doomed to end in the defeat of the conqueror. The Saxon settlers in the United States, Australia, and South Africa could only survive with constant replenishments of Saxon blood from new settlers. Without this outside help the new colony would die out within a century.

It is in the light of this anticolonialism, based on strict adherence to physiological laws of race, that Knox's writings on Jews must be read. To Knox, the Jews were Chaldeans, innate wanderers who had begun to roam the globe long before Abraham. With their original stock still in Chaldea, a branch of the tribe wandered off into Egypt and fused with other races such as the Copts. After their sojourn in Egypt, the Jews dispersed throughout the Near East, first conquering Syria and then fanning out all over the world. This endless movement, according to Knox, has left them "few in number, without a rallying point." (261)

Thus Knox saw the Jews in much the way in which he regarded the Roman, Spanish, British, and French empires—doomed because of their lack of rootedness. Physiological laws dictated that one race cannot colonize another. Knox held that the pages of history were filled with the failures of those empires' efforts to conquer new lands and peoples. The Jews, originally a pastoral people of Chaldean descent, scattered not by an accident of history but by one of race, were dispersed everywhere but were nowhere at home. In the historic period of their wanderings the Jews had failed to gain a permanent foothold, had remained weak, and had persisted only through periodic intermarriage. Knox regarded assimilation as a last-ditch, fruitless effort by contemporary Jews to gain a permanent dwelling-place. By portraying the Jews as a wandering band of colonizers cut off from their life-giving origins, Knox predicted the demographic decline and ultimate disappearance of the Jews. He noted that the Jews' decline in numbers had become apparent even under almost ideal political and social conditions. In Britain, there were less than 40,000 Jews while in France, "with the most unlimited liberty, they amount only to about 70,000" (141). Clearly, the iron laws of race sealed the Jewish fate.

With Robert Knox, British racial science broke away from its environmentalist-humanitarian moorings and headed into the more turbulent seas of innatist intolerance. In making this move, Knox saw fit to do what no British scientist before him had done, namely, to give the Jews a prominent place on deck. As Jews came to enjoy the benefits of political and social changes in Europe, Knox and others came to challenge, through the language of race science, the place and role of Jews in the new Europe, accepting the possibility of cultural, but not racial, assimilation.

Despite the contemporary influence of Knox and the importance of Jews in his anthropological work, their still marginal role in tolerant British society determined that Jews would remain peripheral subjects of British anthropological studies for some time. Indeed, it was eleven years between the publication of the *Races of Men* and the appearance of the next significant English racial study of the Jews. In 1861 John Beddoe delivered an address to the Ethnological Society of London entitled "On the Physical Characteristics of the Jews." It is the first text in English racial science specifically devoted to the physical anthropology of the Jews.

Beddoe continued a trend that began with Knox: the examination of the various "races" of Great Britain and Europe. He was particularly concerned with physical variations in humans, and from the 1850s concentrated on recording the hair, skin color, and cranial formations of the inhabitants of the British Isles. After graduating from the University of Edinburgh medical school in 1853, he served in the Dardanelles as a doctor during the Crimean War. Immediately after the cessation of hostilities he made a grand tour of the Continent, studying medicine and physical anthropology, after which he returned to Bristol in 1857, remaining there in private medical practice until his retirement in 1891. It was during this time that he established himself as a leading figure in British physical anthropology. He was a founding member of the Ethnological Society in 1857, president of the Anthropological Society from 1869 to 1870 and president of the Royal Anthropological society from 1889 to 1891. In delivering the 1905 Huxley lecture entitled "Colour and Race," he was awarded the prestigious Huxley Medal.[47]

The aim of Beddoe's work was more than mere description. He was vehemently opposed to the Knoxian view of "proper places for certain races." He sought to explain human variation in terms of the spread of culture and language, stressing that environmental factors operated in concert with natural selection to produce physical differences. His "racial history" of mankind was inclusive rather than exclusive, environmentalist rather than innatist, and monogenist rather than segregationist. Although he believed in the stability of racial characteristics, he held that they could be explained by climatic influence or demographic intermixture. For Beddoe, Jews were an integral part of the racial history of Europe, and thus there was no question as to their rightfulness of place there.

In "On the Physical Characteristics of the Jews," Beddoe declared his intention to discuss the "peculiarities of form and feature" that make up the Jewish "type." But despite this promise of a broad-ranging discussion, after some brief preliminary remarks on Jewish ophthalmological characteristics and facial structure, Beddoe devoted the rest of the paper to a disquisition on the climate-color correlation as it pertained to Jews. A detailed review of the literature on the subject was followed by an analysis of the data gleaned from Beddoe's own observations of the hair and eye color of 666 Jews of Ashkenazic and Sephardic origin. Beddoe detected two distinct Jewish racial types, both of them within "the Caucasian family, the pure xanthous or rufous and the melanous." Redheaded Jews were to be accounted for not because of the influence of climate, as he thought Prichard suggested, or because of miscegenation, as the French anthropologist Paul Broca advanced, but because red was one of the two original colors of the Jews. The theory held that since Ashkenazim and Sephardim present similar proportions of rufous Jews (the latter have slightly more), Jews in general possessed a trait that somehow is a singular advantage in the struggle for existence. By being composed of these two racial types, the redhead and the brunet, Jews are able to flourish in areas where gentile populations remain stagnant or even decline. Beddoe was of the opinion that "they are able, it seems, to live, thrive and multiply in all countries where any branch of that family can subsist.... In the towns of Algeria they are, according to Boudin, the only race that is able to maintain its numbers, while Frenchmen, Spaniards, Moors and Negroes tend to die off more or less rapidly."[48] Thus Beddoe rejected Knox's hard-line, racially motivated anticolonialism. Rather, Jews became an example of a race that could enter foreign domains as diverse as Sweden and Aden and thrive, often at the expense of the indigenous populations. While the roots of this notion are set deepest in Germany and Austria, isolated concerns about Jewish competitiveness began to be voiced in England from the time of Beddoe.

Unlike in Central Europe, England's fears were not based on a perception that Jews were overrepresented in the cultural life of the nation to the point where they begin to dominate. Rather, they were bound up with intimate concerns over Britain's "racial fitness" or what was known as "National Efficiency." The India mutiny (1857), the Jamaica revolt (1865), and the poor showing in the Boer War (1899–1902) combined to launch a panic among the English that the sun was about to set on their empire and that radical racial improvement was necessary to stop the decline.[49] In this atmosphere of national humiliation and uncertainty about the future, as well as large-scale East European Jewish immigration to Britain after 1881, a ground swell of anti-Jewish sentiment began to rise in England.[50] However, although Jews were lampooned in the popular press and even made the focus of attack by anti-immigrationists, they never became the targets of a sustained campaign of hate such as was directed against them on the Continent. Indeed, it may be taken as a measure of the extent and nature of their integration into the body politic that Jews continued to be overlooked in British anthropological literature even after this crucial historical juncture.

Nonetheless, it was partially because of the change in public sentiment in England, and primarily because of the overt antisemitic campaign on the Continent which found support in scientific racism, that in 1885 a young Anglo-Jewish scholar, Joseph Jacobs, sought to wrest control of the anthropological discourse on Jews away from gentile scientists. In so doing, Jacobs redefined established scientific paradigms and created Jewish racial science.

FOUR

JOSEPH JACOBS AND THE BIRTH OF JEWISH RACE SCIENCE

> The Jews have been made what they are by the Bible.... Their life has been dominated by its law, their feelings by its psalter, their ideas by its prophets, their outlook on life by its wisdom, and their hopes for the future by its apocalypse.
> *Joseph Jacobs, "The People of the Book"*

Joseph Jacobs was the first of the Jewish race scientists, and one of the few people in Victorian England to examine the Jews as a race. Because of the racial makeup of the British empire and the success of Jewish integration into British society, the racial characteristics of Jews were of minor scientific interest to British anthropologists. Eschewing academic convention of the time, Jacobs focused anthropological attention on Jews in a way that no one in Europe had done before him. Moreover, because his methodology was primarily statistical and comparative, it marked a significant shift away from the earlier descriptive efforts of travelers and missionaries.

Jacobs's efforts were very influential on the Continent, especially in Germany, the country that became the focal point of the new discourse on the biological history of the Jews. His work was also in large part motivated by the virulent antisemitic movement in Germany in the last quarter of the nineteenth century, and many of his studies were actually based on statistical and anthropological evidence gathered from the

German empire. His findings were referred to by almost every author who addressed the problem of Jewish physical anthropology, not only in Britain and on the Continent but throughout the world.

Jacobs was one of the earliest outspoken defenders of the concept of a single, pure Jewish race. The question of racial purity was central to scientific racism. Even in the face of criticism from other physical anthropologists, many of them Jewish, he staunchly defended this view. Jacobs's argument was founded upon his reading of Jewish history, using original Jewish and Christian sources, and anthropometry, based on data he had gathered personally. The result was a radical interpretation of the Jewish past and present that acknowledged the intimate connection of history and biology. In approaching the problem of the anthropology of the Jews from a sociological and historical point of view, Jacobs equipped himself with a corpus of responses to those critics of the Jews who saw many of their most hated characteristics not as historically determined but as biologically inherent. In the process, he showed that a belief in racial purity did not have to mean a belief in racial superiority. His unshakable faith in the power of environment over "germ-plasm" meant that his anthropology was informed by high humanistic ideals and not violent jingoism.

Jacobs challenged the view that held the Jews to be physically and mentally degenerate and Judaism a fossil religion. These opinions prevailed during the European Enlightenment and continued to strike a resonant chord among antisemites of the late nineteenth century.[1] Inspired by the nationalistically charged atmosphere of the time, Jacobs's parochial anthropology constituted a new form of Jewish self-defense. With his categoric rejection of both the anti-Jewish trend of the Enlightenment and the racial antisemites of his own day, he became one of the most eloquent and intellectually sophisticated defenders of the Jewish people. Scientific resistance to stereotypes of inferiority was unique to minority groups, and Jacobs was one of the earliest exponents of this type of response.

Jacobs's work covered the gamut of racial science, and he asked questions that were crucial to all its practitioners. What was the role of the environment in connection with race? Could a changed environment change a race? Were certain races susceptible to certain diseases? And what were the anthropological characteristics that set one race (in this case the Jews) apart from others? What role does head form play in determining the relative purity of a race? Can the findings of modern archeology help set the anthropological record straight? How can one incorporate the new science of photography into modern anthropological studies? What was the role of statistics in race science, and could they be of use in solving the Jewish problem?

As science is contingent upon culture, and is not created and does not exist in a value-free environment, the historian of a particular episode in the history of science would do well to know something of the scientist's biography. The life histories of the Jewish race scientists had direct bearing on the science they produced. In Joseph Jacobs's case, to be familiar with his own personal journey, intellectually and physically, is crucial to appreciating the scope of his tremendous literary output, but especially his race science.[2] When Jacobs was born in Australia in 1854, the country was in the midst of an enormous gold rush. The chance to strike it rich led to a large-scale immigration and the rise of the two major urban centers, Melbourne and Sydney. Among the newcomers were Jews. At mid-century the Jewish population of Australia stood at about eighteen hundred, and by the time of Jacobs's birth in Sydney, the number of Jews on the continent had reached five thousand.[3]

Almost nothing is known of Jacobs's early Jewish education other than that he was an honorary teacher at the Sydney Jewish Sabbath School in 1871. He then spent one year at Sydney University, in 1872, where he won class prizes in classics, mathematics, chemistry, and experimental physics, disciplines that prepared him well for his later anthropological work. But Australia was an isolated place, unable to satisfy Jacobs's intellectual appetite. It is little wonder that, like many others in his position, his ambitions steered him towards England, and sometime in 1872 he left Australia for Cambridge, never to return.[4] He arrived in England with letters of recommendation to the family of the Anglo-Jewish historian Lucien Wolf, with whom he eventually made a lifelong friendship and close working relationship.[5] Jacobs enjoyed a distinguished academic career at Cambridge; he was awarded first class honors in the Moral Science Tripos of 1876 and named Senior Moralist for that same year.[6]

The year 1876 was not only one of public recognition for Jacobs but was also one of profound personal discovery. According to his own testimony, it was the appearance that year of George Eliot's *Daniel Deronda* that fixed him on the course of Jewish scholarship. Jacobs later recalled that "when it appeared I was just at that stage in the intellectual development of every Jew, I suppose, when he emerges from the Ghetto, both social and intellectual, in which he was brought up.... George Eliot's influence on me counterbalanced that of Spinoza, by directing my attention, henceforth, to the historic development of Judaism. Spinoza envisaged for me the Jewish ideals in their static form, George Eliot transferred my attention to them in their dynamic development. Henceforth I turned to Jewish history as the key to the Jewish problem."[7]

Jacobs saw in Eliot's version of Jewish history a literary analogue to contemporary scientific thought. He firmly believed that Jewish history, with its gradual evolutionary development, was an example of applied Darwinism, because over time those qualities that assisted it in the struggle for existence were preserved and those that did not were discarded. The importance of Darwinian theory to Jacobs and the intellectual vistas it opened up for those of his generation is made clear in his appreciation of the Darwinian quality of Eliot's work: "It is difficult for those who have not lived through it to understand the influence that George Eliot had upon those of us who came to our intellectual majority in the 'seventies.' Darwinism was in the air, and promised, in the suave accents of Professor Huxley and in the more strident voice of Professor Clifford, to solve all the problems of humanity. George Eliot's novels were regarded by us not so much as novels, but rather as applications of Darwinism to life and art. They were to us Tendenz-Romane, and we studied them as much for the Tendenz as for the Roman."[8] As Darwinism promised to unveil the mystery of the development of life, so too a Darwinian approach to Jewish history would unearth the key to the development of Western civilization, the Jewish contribution to that development, and a solution to the Jewish problem. For Jacobs, his assiduous pursuit of these goals was to play a large part in his battle against Continental antisemitism.

Jacobs alone among the Jewish race scientists was not prompted to develop his anthropology because of a personal encounter with antisemitism. England had been for him, as he hoped it would be for other Jewish immigrants irrespective of their countries of origin, a safe haven.[9] For Jacobs, the Jewish problem was most clearly manifest in the Russian pogroms of 1881 and the discriminatory practices that still obtained in Prussia and the Austro-Hungarian empire. Against these he launched into combat, attempting to defend the Jews in a rational, dignified, and novel way by appropriating the discourse of modern science, in this case physical anthropology, and combining it with a sociological interpretation of Jewish history.[10]

Inspired by Eliot and energized by Darwin, Jacobs set out to acquire the specialized training necessary for his self-appointed task. To place his Jewish education on firmer footing, he spent 1877 in Germany studying Jewish history and texts with the distinguished scholars Moritz Steinschneider and Moritz Lazarus. Jacobs's view of Jewish history was integral to his anthropological writings on the Jews. Although he never resorted to a crude Gobineauian reading of the centrality of race in the unfolding of the historical process, he was convinced that certain races and nations were unique, and that these racial differences, which revealed themselves in the intellectual output of a race, were decisive. Jacobs felt that a study of Jewish history, when combined with an analysis of Jewish racial characteristics, would provide him with a powerful arsenal in the battle against the antisemites and therefore with a possible solution to the Jewish question. He regarded it as his duty to fight the antisemites of his day by pointing out Jewish "contributions to civilization." This required a dual approach that entailed both historical investigation and quantitative analysis. If one could actually prove by example, and then statistically measure the worth or extent of Jewish contributions, then the empty claims of the antisemites could be systematically and persuasively rebuffed.[11]

Upon his return to England in 1878, Jacobs sought to combine his Jewish studies with those of modern science, studying anthropology with the noted statistician and scientist Francis Galton. This eventually blossomed into a rich working relationship between the two men. Having sought out and been taught by some of the most preeminent scholars in their fields, Jacobs had by now acquired a solid foundation upon which to undertake his studies of Jewish history and Jewish racial characteristics. In both areas he was regarded as nothing less than a pioneer.[12]

As an immigrant, Jacobs's perspective on England, particularly Jewish life there, was always that of the outsider. To be sure, coming from Australia he required minimal cultural adjustment to fit into British society and had an advantage over the thousands of primarily Russian Jewish immigrants who arrived in Britain between 1881 and 1914. They, who had arrived in England under very different and far less privileged circumstances, fled grinding poverty and vicious pogroms, while Jacobs had come at leisure to study at one of the world's most prestigious institutions of higher learning. Nevertheless, because he was an immigrant, he was a keen and sympathetic observer of this massive wave of migration and became actively involved in the successful absorption of the Jews. He was prompted to do so, not only because of what was described by all who knew him as his loving humanitarianism, but also because unlike his own, the immigrants' road to becoming proper English men and women was extremely rocky. He saw it as within his power to clear the path for them.

In addition to his actively assisting Russian Jews upon their arrival in England, it was Jacobs who first alerted the world to their plight and suffering, in two articles that appeared in the *Times* on January 11 and 13, 1882. As a direct result of Jacobs's public outcry, a great meeting to protest the persecution of the Jews in Russia following the assassination of Czar Alexander II was held at the Mansion House on February 1, 1882. Attended by major figures in Britain's public and intellectual life, the meeting was supported by, among others, the Archbishop of Canterbury, Cardinal Manning, Charles Darwin, Matthew Arnold, Lord Tennyson, and Lord Shaftesbury. Aside from a strongly worded condemnation of the pogroms, the meeting raised the substantial sum of £108,000 to aid the victims. Jacobs also served as honorary secretary of the Mansion House Fund and Committee.[13]

The arrival of so many Russian Jews to England occasioned Jacobs's major studies of the Jews as a race. Certainly, by having studied with Charles Darwin's cousin Galton in 1878, he had already demonstrated an interest in the subject. But it was the rise of political antisemitism in Germany, personally observed during his own sojourn there, and its more violent manifestation in the Russian empire, that led him to begin collecting his own anthropological statistics, on which basis Jacobs created the new field of Jewish racial science. In the preface to his most important collection of racial studies, *Jewish Statistics* (1891), Jacobs informed the reader that "the following studies began in an attempt to get reliable data about the Jews of Europe when the antisemitic movement was at its height."[14] Jacobs's remark indicates that the study of race in the modern period was always more than pure science. Whether in action or reaction, it was often either motivated by or soon annexed to political causes.

Aside from his concern for the welfare of Russian immigrants, Jacobs was at pains to show that Jews in general were an assimilable group. In the light of the massive population shifts taking place, Jacobs's studies were designed to allay the fears not only of the British government but also of other countries now confronted with great numbers of Jewish refugees. Wherever they went, he maintained, Jews soon participated to the fullest extent in the culture of their adoptive lands, displaying diligence, sobriety, and thriftiness. These cherished bourgeois values were therefore as integral to Judaism as they were to the modern European nation-state. In what was a sociological variant of Moses Mendelssohn's axiom of Judaism and Christianity sharing a common canon of natural law, Jacobs saw modern Jews and Christians as sharing a similar value system and life-style.

According to him, these qualities of the Jewish character were social and not biological in origin, and thus the evolution of them could only be understood by a study of the history of the Jewish people. In fact, Jacobs maintained that the study of Jewish history and the physical and intellectual properties of its actors would be of inestimable value for the general study of history and anthropology. His faith in the universally instructive potential afforded by a study of Jewish life, both historical and biological, was bound to his belief in the historic "mission of Israel."[15] Jacobs sincerely believed in the moral imperative of the Jewish people and that Jewish history was purposeful in that it breathed meaning into the historical course taken by other nations. In an address to the Royal Academy of History in Madrid in 1895, entitled "Jewish History: Its Aims and Methods," he declared that "it is the conviction of many others, besides us Jews, that a Divine purpose runs through the long travail of Israel. Jews alone form a bridge between ancient and modern times. If their history does not contain any inner meaning, then the life of man upon this earth has no rational aim."[16]

As part of his holistic philosophy, Jacobs also believed that the anthropological study of the Jews would shed light on the problems encountered by all anthropologists in studying race: "I do not think we Jews quite appreciate the immense interest of the study of ourselves, merely regarded as an intellectual pursuit. Beginning with our very physique, we plunge at once into all the intricate problems of heredity, and the even more intricate problem of the influence of the social conditions on the bodily organization. A man might get himself no inconsiderable reputation in merely devoting himself to the study of Jewish biostatics. Then again, our hygiene brings the student of Jewish affairs in contact with some of the most interesting of medical problems; while our vital and other statistics have their instruction to give to the growing science of vital statistics."[17] So the spiritual makeup combines with the physical nature of the Jews to become a vehicle for all peoples to view and appraise themselves.

Jacobs's first systematic writings on the Jews appeared in as a series of articles published in the *Jewish Chronicle* between 1882 and 1885.[18] These seven "Studies in Jewish Statistics" were the result of one immensely productive year he spent amassing all manner of statistical data on the Jews of England and the Continent. The thrust of these essays was sociological and should be seen as the historical and sociological portion of his broader project: the writing of a biological history of the Jews. They stand, in fact, as a preface to his later anthropometric writings on race science. For Jacobs, Jewish history and anthropology were inseparable. His intent was to study various aspects of Jewish life in order to ascertain which of those characteristics were the product of environment and which were biologically determined. As we will see, Jacobs was never a crude racial determinist, and much of his anthropological work was taken up asserting the primacy of nurture over nature.

The essays, the first works of their kind, are remarkably complex and detailed, a fact for which Jacobs felt obliged to apologize on a number of occasions.[19] In Great Britain, statistically based sociological surveys of various groups in the population were extremely difficult to undertake because the census did not classify inhabitants according to religion. Jacobs was thus forced to seek out other sources, and eventually he accumulated an enormous quantity of data from many places, including the Board of Guardians, burial societies, synagogues, and Jewish hospitals and schools. As a result, he was able to provide the most comprehensive, statistical portrait of Anglo-Jewish life available up to that time. His aim was to dispel myth, establish fact based on mathematical evidence, and defend Jewry against the charges of German antisemites and anti-immigration circles in Britain.[20]

Originally published in serial form, Jacobs's seven studies complement each other, and together they create a comprehensive picture of the contemporary Jewish life cycle. In the first essay, "Consanguineous Marriages," Jacobs went straight to the heart of an issue that had long plagued European politicians and now the medical profession: the nature of Jewish marital culture. Seventy-five years earlier, when Napoleon convened the Jewish Sanhedrin of 1807, he asked as the first of twelve questions about Jewish ritual life and belief, "Are Jews allowed to marry more than one wife?" Although the question did not directly address the topic of consanguineous marriages, its sexual nature points to the emperor's belief in the supposed immorality of the Jews.[21] More recently, the alleged evil consequences of first-cousin marriages, and Jewish intramarriage in general, had become a hotly debated topic in anthropological and medical circles. Jacobs thus approached the subject with the twin themes of alleged Jewish immorality and degeneracy in mind.

Jacobs's aim was twofold. He first sought to determine whether there was indeed a higher proportion of consanguineous marriages among Jews than non-Jews, and then whether a causal link could be drawn between these first-cousin marriages and the seemingly higher incidence of mental and physical diseases among the Jews. In adopting the method devised by Sir George H. Darwin, Charles's son, to determine the extent of first-cousin marriages among English Christians, Jacobs proceeded to do the same for English Jews. He did this by examining all the marriages that were announced in the *Jewish Chronicle* from 1869 to 1882 and concluded that 7.5 percent of all marriages between English Jews involved first-cousins. By comparison, only 2 percent of marriages among English Christians were consanguineous.[22]

Jacobs gave as the historical reasons for the prevalence of first-cousin marriage among Jews the absence of theological proscriptions against the practice (such as obtained in the Catholic church), "the existence of small communities scattered about, the rare communion between the sexes, and, above all, the absence of any ideal of pre-nuptial love." He adduced several other reasons for its prevalence in England, whose roots lay in the particular historical conditions under which English Jewry lived. These were the absence of a *shadkhan* (marriage broker), who would have brought people together from different parts of the country; the relative wealth of English Jews, which led the wealthier families to marry among themselves; and the practice of what he termed "shoolism" or limiting one's circle of friends and acquaintances to one's own synagogue.[23]

The statistical and the historical aspects of his study now complete, Jacobs moved onto the anthropological or medical side of the question, namely, are there any ill effects caused by consanguineous marriage? The medical world was split on the issue. From about 1860 Darwin's conclusions about the beneficial effects of cross-fertilization, and the injurious ones of self-fertilization, greatly influenced those who held that marriage among kin was medically harmful.[24]

Opponents of inbreeding claimed that the practice caused conditions such as diabetes and deaf-mutism, as well as various forms of mental illness. The Jews were singled out by contemporary anthropology and medicine as the quintessential victims of such marriage patterns. Those who believed in the harmfulness of consanguineous marriages called into question the ability of such unions to produce offspring. It had been alleged that they were sterile, but Jacobs's investigations of sixty-two consanguineous marriages found only three to be barren. He calculated the proportion of sterile marriages among Jews who had married their first cousins to be 5.4 percent, a figure that compared very favorably with the national average of 16.3 percent. Furthermore, Jacobs found such marriages to be especially fertile, producing 4.6 children, as compared with the English average of 2.26.[25]

Jacobs thus rejected the idea that consanguineous marriage was harmful to the Jews, and in so doing concurred with several other distinguished researchers on the subject.[26] He cited with approval the findings of one contemporary author who held that rather than cause disease, consanguineous marriage had actually led to the "superior vitality and cosmopolitanism of Jews." It was this aspect of Jewish marriage that provided them with the biological mettle to withstand, indeed emerge triumphant from, the deplorable conditions of the ghetto.[27] Jacobs, as well as a number of later Jewish anthropologists, argued that ghetto conditions had been the mechanism of a process of selection whereby only the fittest survived, and that endogamous marriage served to preserve those vital characteristics. In this respect, the entire course of Jewish history, even if lachrymose, was purposeful, and indeed given existential meaning and biological justification.

Jacobs was not content with this conclusion, apologizing "to the reader for leading him over so arduous a path to so unsatisfactory a goal." Yet it is clear that Jacobs did achieve some success. His accumulation of hitherto unknown facts and figures about English Jews was extremely valuable in that it gave a clearer picture of the marriage patterns of this community in the late nineteenth century. His dismissal of the claims of the ill-effects of consanguineous marriage, and his presentation of evidence to suggest the great fecundity of such unions, was a challenge to those who maintained that the prevalence of mentally ill and physically handicapped offspring among European Jews was a direct consequence of their immoral way of life. Jacobs was a staunch defender of Jewish morality, and his firm belief in the historic mission of Israel was grounded in his faith in the permanency and universality of the moral edifice that Jewish civilization had built.[28]

Jacobs continued to address the theme of Jewish morality in the second of the seven essays, "The Social Condition of London Jewry." A stock-in-trade cry of the antisemites was that Jewish civilization was immoral, in great part owing to what was seen as its pathological obsession with money and the attainment of riches. The idea had tenaciously survived since Rome and had gained universal acceptance: from the New Testament to Shakespeare to Marx to the czarist secret police, antisemitic invective was heaped upon the Jews for supposedly worshiping mammon. This belief, and the concomitant notion that all Jews are wealthy, became the focus of Jacobs's statistical study of the social condition of the Jews of London.

His aim was to defend Jews against the charge of excessive wealth by pointing out, with the aid of statistics, the vast number of poverty-stricken among them, and to show that the occupational structure of Jews was a product of specific historical conditions. Jacobs sought to challenge the widely held notion that Jews lived where they did and earned a living as they did because of racial peculiarities.

Jacobs maintained that the poverty of the Jews was generally not recognized because most Europeans assumed that since the Jews they came into contact with in the cities earned a living mainly through commercial pursuits or financial transactions, all Jews must be similarly employed. But the main cause for this misconception was that "the Jewish poor have never been a burden to the general population but have been entirely supported by the Jews themselves."[29] In fact, Jacobs claimed that were the Jews to be regarded as a nation, they would certainly be the poorest among those called civilized. And this applied not only to the penurious Jews of Russia but to those of Western Europe as well.

Jacob's defensive strategy on behalf of Western European Jewry was well crafted to the historical forces in operation at the time he was writing. The last quarter of the nineteenth century was a period of heightened antisemitism in Germany and France. Exacerbating the tensions were a number of financial scandals, some involving Jewish financiers. The ancient accusations of Jewish greed, extraordinary wealth, and aversion to manual labour were coupled with rumors of a world Jewish conspiracy directed by a council of elders.[30] Jacobs had a firsthand view of this growing antisemitic movement when he was studying in Berlin in 1878.[31] And so he chose to focus on Western Europe, with a view to proving not only that wealthy Jewish entrepreneurs were an insignificant minority there, but also that large numbers of Jews received social welfare benefits provided by local Jewish communities.

To make his case, Jacobs employed a comparative approach. He found that in 1861, 6.46 percent of the Jewish population of Prussia was classed as paupers, while the percentage for the whole population of Prussia was 4.19 percent. More extreme was the case of Amsterdam, where in 1877, of the city's 32,500 Jews, 13,000 subsisted on communal charity. Even in a much smaller place, such as the Bavarian town of Fürth, Jacobs noted that a benevolent society gave material assistance to 32,656 Jewish travelers between 1875 and 1879.[32]

Jacobs's presentation of continental statistics served as the background data for the main focus of his investigation: poverty in the London Jewish community. His hope was that a study of London Jewry's financial status would shed light on the prosperity or penury of European Jewry as a whole. For as he saw it, "London is probably the richest Jewish city in the world . . . therefore, if we could calculate the average income of a London Jew, we would know the highest tidemark of prosperity to which

Jews, as a body, have attained." He estimated the average income of a Jew in the English capital to be £82.³³

Of the 46,000 Jews in London in 1882, Jacobs found that a full 11,099, classed as "paupers," received assistance from the city's 47 Jewish charitable institutions. These were maintained by the community at an approximate cost of £37,000 per annum. Another measure of the extent to which poverty reigned among the Jews of England is that in 1882 the Overseers of the Poor of the United Synagogue distributed Passover *matsot* to 2,500 families, or about 10,000 individuals.³⁴ After computing the proportionate distribution of wealth among London Jewry, Jacobs came to three major conclusions: the vast gap between the incomes of upper-class and so-called middle-class London Jews showed that there was no real middle class among them; since one Jew out of every four was a pauper, there was among Jews a similar "intensity of wealth and poverty" as among Christians; and the relatively large number of small merchants, shopkeepers, and professionals, along with the correspondingly few in the "super-pauper" category, "bears eloquent testimony to the industry, prudence, and perseverance of the ordinary Jew. Though he may not make himself rich, he will rarely fail to make himself independent."

Not only was Jacobs statistically able to dispel the myth behind the axiom "rich as a Jew," by taking what he identified as Europe's wealthiest Jewish city and showing the widespread poverty that existed within that community, but he also sought to prove by way of an investigation into the occupational structure of London Jews that the solid bourgeois values of hard work and thriftiness exhibited by all classes of Jews was central to the Jewish value system. Jacobs's aim was to show that Jews did not pursue a different economic agenda (as was often alleged on the Continent) from that of the modern European states in which they lived. Even the masses of poor refugees streaming out of Russia would, because of their Jewishness, soon become good English men and women.

Jacobs's desire that the "foreign contingent," as he called the East European Jewish immigrants, adopt the English way of life was not part of an overall assimilationist philosophy. He cherished Jewish culture in all its manifestations, and rather than seek its dissolution into the body politic of the surrounding nations, he felt that Jewish culture served to fructify and enrich the societies around it. But there was a danger that this may not always be the case. He dwelt at length on what he called "Chinesism," a "great danger that begins to loom before us as never before in the world's history." According to Jacobs, Chinesism "is the tendency among huge masses of people to crush individuality, and to reduce all its members to one dead level of mediocrity." Turning his attention to the Jews, he declared, "suppose for a moment that the whole 100,000 Jews of Great Britain were made, tomorrow, completely indistinguishable from the rest of Englishmen, what advantage would accrue to humanity from the transformation? Is it not possible that the English character might have thereby lost the chance of being enriched at a favourable moment by one or other of the specific Jewish ideals?"[35]

Judaism, and especially Jewishness, were prime values to be retained. What led Jacobs to propose "that the immediate problem before the Jews of London is to Anglicise their 'foreign contingent'" was not a belief in the superiority of English culture but, rather, a health crisis within the Jewish community of London: an appallingly high infant mortality rate, which in 1882, alone among world cities, was higher than that of the general population. Checking the burial statistics of the United Synagogue for 1882, Jacobs found that of the 507 charity funerals for that year, 411, or 81.4 percent, were those of children under ten years of age. By comparison, English and Welsh figures for 1880 showed such deaths tallying 43.5 percent. Jacobs felt the cause of his coreligionists' misery was that they were as yet "unaccustomed to the English climate and conditions of life" in the capital.[36] In making this claim, Jacobs was again arguing for the adaptability of the Jews. They had no racial predisposition to disease, nor did they lead profligate lives that contributed to all manner of illness and to early death. Rather, with nurturing in the right environment, the Jews would thrive and show themselves to be upstanding members of community and country.

But Jacobs's study of Jewish childhood deaths entailed more than this simple hope. His intention in laying bare the "facts" of Jewish life based on statistical evidence was to dispel both stereotype and myth. For example, over the course of centuries, a myth had developed that Jewish parents took better care of their children than non-Jews. By the nineteenth century, this belief, which incorporated notions of a general Jewish immunity to various diseases, found its way into contemporary medical discourse.[37] A similarly long-entrenched myth was that young Christian children were especially likely to die a premature death. In other words, where Christian children seemed inexplicably to succumb, Jewish children tended to survive. In the Middle Ages, Christian society built a psychological crutch to explain the losses of its children by blaming the societal Other, the Jews, whose offspring appeared to survive the trials of childhood. It was believed that since the Jews had killed the son of God, then they were certainly capable, and most probably responsible, for the untimely deaths of innocent young Christian children: hence the persistence of ritual murder accusations.

The irony that, contrary to myth, death visited Jewish children in nineteenth-century London far more frequently than it did gentile ones, despite the tenacity of popular belief, would not have escaped Jacobs, an expert on the history of the ritual murder charge. The historical development of the blood libel, as well as the immediate allegations against the Jews in the Hungarian town of Tiszaeszlar in 1882, were of particular concern to Jacobs.[38] In the midst of working on his statistical studies, he published an article entitled "Historical Notes on the Blood Accusation."[39] Later, in 1894, he would make his most significant contribution to the subject, "Little St. Hugh of Lincoln," the infamous story of a blood libel against the Jews in the English town of Lincoln in 1255.[40] In the merging of history and anthropology, Jacobs made apologetic claims for the Jews, both medieval and contemporary. In the case of the former, he challenged the scanty and often circumstantial evidence, defending the Jews against a most heinous charge; in the latter case, he asserted the humanity and fragility of contemporary Jews, forced by poverty and want to bury so many of their young children.

In the foreground of his vivid statistical picture of contemporary Jewish life Jacobs placed the Jewish worker. Of his seven statistical essays, three—"Occupations," "Occupations of London Jews," and "Professions"—were entirely devoted to the working life of Jews. Influenced by his teacher, Moritz Lazarus, the inventor of the term *Völkerpsychologie* or the psychology of nations, Jacobs maintained that "nothing throws more light on the character of a people than the occupations in which its members pass their lives." He recognized that the accusations of the antisemites about the supposedly limited range of Jewish occupations had to be refuted: "There is no subject relating to Jews on which we know so little; as a matter of course there has been none on which so much has been asserted during the recent antisemitic movement. The importance of the subject, both polemically and as a guide to future improvement, has led me to collect as much as possible relating to it."[41]

The apparent fixity of Jewish occupational activity had excited critical interest for more than a century. Among physiocrats such as Christian Wilhelm Dohm in Prussia, Abbé Gregoire in France, and representatives of both the Enlightenment and its Jewish counterpart, the Haskalah, most commentators, whether friend or foe of the Jews, concluded that the predominance of the petty trader, merchant, and middleman and the absence of agricultural and artisan classes accounted for much that was defective in Jewish society.[42]

Writing at the time of the antisemitic excesses on the Continent, where violent denunciations of the work habits and ethics of the Jews were common, Jacobs was at pains to show that the traditional and stereotypical picture of Jewish occupational structure was an oversimplification.[43] First, as usual, he pointed out sound historical reasons for Jewish occupational choice. Most people in the world, he observed, did the same kind of work as their fathers and forebears, and thus the Jews merely followed a universal cultural pattern. Specifically, however, exclusion from guilds, the denial of the right to own real property, church decrees restricting them to usury, prohibitions against holding land, and residence restrictions, as well as Jewish religious enactments requiring at least ten Jewish males to make up a quorum for daily prayers, explained why since the Middle Ages Jews throughout Europe generally resided in towns and cities and earned a living the way they did. These were, of course, old arguments, and were most fully articulated at least a century earlier by proponents of Jewish emancipation on the Continent.

But Jacobs chose not to dwell on historical prohibitions against Jews doing this or that type of work. Instead he focused on how the exigencies of city living and the requirements of religious practice determined the occupations pursued by the Jews. Like most other contemporary Jews in Western Europe, Jacobs lived in a large, modern city, and his research reflected this fact. Indeed, he was the first Jewish researcher to ask about the effects of what the Jewish historian Salo Baron later called the "metropolitanization" of the Jews. What is more, Jacobs's work significantly predates what is generally referred to as urban sociology, as well as Max Weber's religious sociology in the early twentieth century. Although his analysis may seem obvious today, given the time and circumstances in which it was written Jacobs's reading of the situation was rather sophisticated. According to him, occupations were chosen because expanding cities required the services that Jews, and increasing numbers of non-Jews, could provide. Considering that until 1914 the United Kingdom was the world's main trading nation and London the world's financial center, it is to be expected to find the English, both Jews and gentiles, involved in a wide variety of modern commercial and industrial enterprises. The same was also true for other European capitals. For as Jacobs remarked, "it is absurd to expect a man who lives in Paris, Vienna or Berlin to tend sheep or dig up coal." Nevertheless, Jews were denounced for the narrow spectrum of their occupational choices, in particular the absence of farming.

But with the universal movement to the cities in the late nineteenth century, Jacobs correctly noted that to examine the way Jews earned a livelihood "we must compare Jewish occupations with those of the town dwellers."[44]

When he did this, he found that the occupations of the Jews were extremely varied. Although pockets of concentration, such as tailoring, did exist, the number of Jews working on the stock exchange, for example, or in key industries was greatly exaggerated.[45] Conversely, Jacobs found that the number of Jews engaged in handicrafts was always underestimated by antisemites, "who very probably never handled a hoe or a hammer in their life," as well as by Jewish advocates of occupational reform. In fact, when Jacobs examined the occupations of London Jews, he found that the Lads' Institute, a branch of the Jewish Working Men's Club, was training 347 Jewish boys in 91 different trades. He took this as a testament to the occupational versatility of London Jewry and as a clear sign that English Jews were in a superior position to their Continental coreligionists, who were still subject to certain occupational restrictions.[46]

For Jacobs, Judaism itself was to a great extent responsible for the occupations pursued by Jews. But even with its stringent demands on the individual, such as dietary restrictions and prohibition against work on the Sabbath and religious holidays, Judaism did not prevent its practitioners from taking up useful and productive labor. Thus for Jacobs, Jews could be both faithful to their heritage and a benefit to their country. The religious requirements concerning the proper slaughtering of animals created a need for skilled butchers; reverence for books and scholarship made Jewish printers and binders among the finest in Europe; and since Jews were permitted to sell their wares on Sunday, Jewish fruiterers and tobacconists could provide a service to the gentile community. "As a general principle," Jacobs concluded, "those trades are most favoured by Jews which afford them opportunities for arranging their own time for work, and leav[e] them free for their festivals and religious duties generally. Piece-work rather than time-work, domestic industries rather than factory work, in fact occupations in which they can be, to a certain extent, masters, would naturally be chosen by a people whose holydays differ from those of their neighbours. Add to this certain natural tendencies, heightened by historic causes, towards banking and international exchange, and the chief occupations of the Jewish race are accounted for."[47]

Aside from attempting to disarm the antisemites with studies demonstrating the "productive" labor of the Jews, Jacobs's essay "Professions" aimed at determining the Jews' "intellectual calibre and special capacities" and so laid the groundwork for his soon-to-be published work on Jewish physical anthropology.[48] The issue of innate intellectual capacities (thought to be measurable) and the notion that each people possessed a unique "racial soul" (unmeasurable) went to the very heart of race science, and the belief that certain races had peculiar talents and deficiencies was the cement that held the unstable house of racial science together. For when quantification failed, anthropologists resorted to the unquantifiable concept of innate tendencies or creative powers to bolster their arguments about the superiority or inferiority of a nation.[49] In this respect, Jacobs's mathematical attempts to come to an understanding of the "intellectual calibre and special capacities" of the Jews bears the stamp of Francis Galton, whose famous *Hereditary Genius* (1869) was an attempt to quantify intelligence and prove that it was passed from generation to generation. Moreover, this idea was directly inspired by his interest in race science: "The idea of investigating the subject of hereditary genius," wrote Galton, "occurred to me during the course of a purely ethnological inquiry, into the mental peculiarities of different races."[50] In following Galton's lead, Jacobs was operating squarely within the mainstream of late-nineteenth century anthropology.

Jacobs's emphasis on the intellect of the Jews addressed an issue that was central to race science at that time: skull measurement. Race scientists sought to determine what effect professional activity, or as Jacobs called it "brain work," had on brain size and cranial dimensions. The issue was of special importance where Jews were involved, because even though they were generally considered to be a racially bonded group, they nevertheless displayed heterogeneous craniometric characteristics. Since skull size and shape were thought to be relatively stable, differences within one population would raise the question of the homogeneity and purity of that racial type.

One point of Jacobs's study of Jewish professions was purely apologetic. Antisemitic movements on the Continent charged Jews with destroying the national Christian character of their respective countries by playing too large a role in the professional and cultural life of the nation. Jacobs did not dispute the percentages the antisemites gave for Jewish involvement in various spheres of professional activity, but he challenged their interpretations of them. Jews had been accused of infiltrating the cultural life of the nation with aims of subverting it. They were regarded as musical, artistic, and linguistic mimics, devoid of any innate creative powers.[51] In his essay on professions Jacobs rejected this thesis out of hand, and continued to refine his argument in three later works: "The

Comparative Distribution of Jewish Ability" (1886), *Men of Distinction* (1916), and *Jewish Contributions to Civilization* (1919). In all, he attempted to prove the existence of what he called "Jewish genius" and of the debt that Christian civilization owed to it. Jacobs's findings on Jewish professional life showed that Jews were "largely represented in the principal professions in all Continental countries outside Russia."[52] In law, medicine, literature, journalism, art, politics, academia, and chess (which Jacobs explicitly classed among the professions)—everywhere except the clergy and the military—Jacobs found Jews in numbers comparable to or even exceeding their percentage in the general population.

From this statistical study of Jewish intellectual elites, Jacobs concluded his picture of the sociological character of contemporary European Jewry with an examination of Jewish "Vital Statistics," an essay on the marriage, birth, and death patterns of the Jews. The "vitality" of a nation—the number and sex of its children, the fertility of its women, and the health and longevity of its people—was another central concern of race science. In Jacobs, the influence of the National Efficiency movement in England is apparent. National Efficiency was not a uniform, homogeneous ideology; it accommodated concepts as far-ranging as tariff reform and eugenics. But all advocates of National Efficiency saw it as a program designed to ensure Britain's competitive edge in every field of human endeavor. At the very least, such an undertaking required a healthy, virile population, and therefore the physical and mental condition of the "race" preoccupied proponents of the movement.[53] Jacobs held that an examination of the biostatistical data relating to the Jewish life cycle would provide clear sign posts for the future development of the race. Central to any such examination was marriage, for Jacobs held that "it is upon the number and nature of marriages entered upon by a people that the whole 'movement' of population depends." When discussing movement, Jacobs meant the consequence of the age at which marriage is entered into, because that determined such things as the average duration of a generation, fertility and fecundity, the overall health of the children, and even a child's sex. In all these categories Jacobs noted that Jews differed from non-Jews.

It was widely held in the nineteenth century that since the Jewish birthrate was lower than that of non-Jews, Jews married less frequently than others. Indeed, in Austria, for example, in 1870 there were 53 Jewish marriages per thousand compared to 98 for Christians; in France (1855–59), 62 to 82; in Hungary (1864–73), 64 to 105; in Prussia (1820–76), 75 to 88; and in Russia (1867), 87 to 100. Generally accepted explanations for these figures were the higher percentage of town dwellers among the Jews (people who were less likely to marry than country dwellers), the larger number of Jewish women than Jewish men in most parts of Western Europe, and the greater poverty among the Jews. But Jacobs was unsatisfied with these explanations. Rather, in his opinion, the reason that Jews married less often than non-Jews was that they had more children per marriage than gentiles, thus reducing the number of marriages. Therefore, Jews did not fit the universal pattern which held that there was a correlation between marriages and births.[54] Since Jewish race science, like non-Jewish race science, was conditioned by the politics of the individual scientist, a set of statistics could be interpreted in very different ways. For example, the Zionist race scientists who followed Jacobs accepted the lower incidence of marriages among Jews as a fact, recognizing assimilation as the casue. This occupied a central position in their own anthropological work on the Jews.

The single most important fact concerning marriage, according to Jacobs, is the age at which it is entered into, because it affects "the physical, mental and social traits of a people." Therefore, it was crucial to determine the proportion of either sex marrying before the age of 20 and between the ages of 20 and 30. Basing his findings only on figures pertaining to Eastern and East Central Europe, he calculated the proportion of Jewish marriages taking place when both parties were under the age of 30 to be greater than it was among Christians, in all cases except Posen. Thus, it remained for Jacobs to determine both the cause of early marriage among Jews and its effect on Jewish offspring.[55] In other words, did early marriage among Jews hinder the attainment of Jewish National Efficiency?

Jacobs proffered several reasons for the early marriages of Jews. These could be divided into two broad areas: physiological and cultural. Of the physical causes, Jacobs cited the greater proportion of women to men among the Jews, which exceeded the proportion of females to males in the general population. This preponderance of young women, when coupled with the massive westward migration of thousands of young Jewish men, led to a natural competition among the women for the limited number of eligible bachelors. A woman who did not marry early was likely never to marry at all.

Another physiological reason, according to Jacobs, was the early appearance of menstruation among Jewish females. The study of comparative menstruation among different peoples was of great concern to physiologists at the time Jacobs was writing. Contemporary studies determined that the age when menstruation first occurred depended on such factors as climate, occupation, and town residence. Racial origin, however, was deemed the most important factor in determining the date of the first appearance of the menses. Nearly all such studies found that Jewish girls menstruated earlier than non-Jewish ones. Most said that Jewish girls menstruated on average at fourteen years of age, while one researcher's survey of a hundred Jewish and Slavic girls discovered the Jewish ones to be menstruating at thirteen and only one of the gentile girls to be menstruating at so young an age. On the basis of such evidence, Jacobs concluded that early menstruation among Jewish girls was a racial trait and therefore an impetus to early marriage.[56]

But in Jacobs's view the most prominent causes of early marriage, especially in Eastern Europe, were social and religious. Jewish males in Eastern Europe tended to marry earlier than their coreligionists in the West, who postponed marriage until they had achieved a measure of financial security. The abysmal economic state in which East European Jewry lived, conditions affording little hope for the future, obviated the desire to delay marriage in the hope of financial betterment. Furthermore, rabbinical authority and talmudic prescription, which bore greater weight among East European Jews than those in the West, served to promote early marriage among the former.[57]

Thus, although Jacobs identified the causes for early marriage in the East to be primarily cultural, he clearly determined that the effects were physiologically or, in the National Efficiency sense, racially detrimental. He found that the custom of early marriage and the practices associated with it were morally debasing, declaring them a threat to the welfare of the people. In Darwinian language, Jacobs declared: "It is easy to see how early marriage handicaps Jews in the struggle for existence by causing them to give, in Bacon's phrase, 'hostages to fortune.'" This flagging effort in life's struggle was, according to Jacobs, partly responsible for the poor physical development of the Jews, "account[ing] in some measure for [their] small height and girth."[58] Jacobs also deemed the institution of early marriage to be psychologically damaging and morally pernicious. He decried the absence of prenuptial love and feared that early marriage "tends to cut short the experiences of the bride and prevents her becoming a helpmate meet for her husband, narrows her views of life and contracts the horizon of her interests to her own household."[59] Worst of all, the custom of the *shadkhan*, or marriage broker, helped reduce the blessed event to a mere business transaction.

In this process of cultural distancing, Jacobs drew a sharp distinction between Continental Jewry, especially East European, and Anglo-Jewry.

The Jews of England were, for Jacobs, the model of modern Jewry, and he the model Jewish immigrant. Therefore, Anglo-Jewry's customs and habits should become the norm for the masses of East European Jewish immigrants, just as they had been for him. In this respect, he seemed to recommend the same kind imperial "civilizing mission" as England had undertaken in its multiracial empire. Almost always, the enterprise and language of Jewish race scientists removed them from their subjects. This is particularly clear with the Zionist scientists whose reformative political agenda included the use of medical metaphors such as "sick" and "healthy" to describe one group of Jews or another. Jewish race scientists, then, seemed to stand apart from their people. This is paradoxical, because to a great extent the Jewish scientists had taken up race science as a response to the antisemitic character of contemporary medicine and anthropology. Obviously, such discourse did not distinguish between Jewish scientists and the Jewish masses.

The study of marriages raised the problem of births. Frequency (births per thousand), fecundity (births per marriage), sex of offspring, illegitimacy, and stillbirths were paramount issues in determining the "viability of the race," to use the language of the day. Declining birthrates and the specter of racial debilitation were of paramount importance in all European nations at the fin de siècle. A large and fit population was taken as a sign of national health and vitality. Besides its symbolic value, it was a key to achieving prestige on the international stage because of the economic and military clout a country possesses when it boasts a large citizenry. Where did the Jews fit into these trends?

Jacobs argued that despite lower pregnancy and childbirth rates, the Jewish population was growing faster than neighboring populations—Jews had larger families, that is, bore more children and raised them to adulthood—because of low infant and early-childhood mortality. He declared (correctly) that the Jewish population was everywhere increasing at a higher rate than that of their neighbors, and that it was the lower number of deaths under five years of age among Jews that made it appear that they were having fewer children. Jewish women, he argued, had fewer pregnancies because the survival rate of their offspring was higher than that of surrounding populations. In fact, the fertility of European Jews in the nineteenth century was immediately recognized by some in the medical community. A Professor Hardy, chief of the medical clinic at the Hopital de la Charité in Paris, shocked his audience at a meeting convened to discuss the alarmingly low birthrate in France by suggesting that all previous attempts to rectify the problem were futile and bound to fail. He proposed that gaps in the population be filled by large-scale Jewish immigration. The Jews, he declared, "are intelligent, industrious, and honest, and what is of greater importance, fruitful."[60] This interpretation was vigorously challenged by later researchers, especially Zionists such as Felix Theilhaber and Arthur Ruppin, who sounded the alarm that the desperately low Jewish birthrate created the very real danger that the Jews would die out.[61]

But Jacobs preceded these two men by almost a generation, and in his time Jewish biostatistics indicated that the Jewish people were healthy, vital, and had a promising future. Only the grinding poverty of the majority of Jews held them back. That the notion of racial extinction was not within the realm of possibility for Jacobs was confirmed by his research on Jewish fecundity. Nearly all indicators showed that "the average number of children to a Jewish marriage is almost invariably greater than those who fall to the lot of Christians." Jacobs suggested that this was the natural consequence of early marriage among Jewish women, the lower rate of stillbirths among Jews, and the higher rate of first-cousin marriages, which were unusually fertile.[62]

Jacobs's most intriguing discovery was that fewer children were produced in mixed marriages than in endogamous ones. Looking at the figures accumulated by Arnold von Fircks for Prussia between 1875 and 1881, Jacobs found that Protestant marriages produced an average of 4.3 children, Catholic ones 5.2, and marriages between Jews and Christians 1.7.[63] Scientists were well aware that individuals of different species did not usually cross, and when they did, produced none or only infertile offspring. For polygenists, the human being was a troubling anomaly. If the human races were in fact different species, originating from separate ancestors, how is it that they could cross to produce fertile offspring? Jacobs was not a polygenist, but he came perilously close to employing a polygenist argument of species reproduction to resolve the issue of offspring in Jewish-gentile marriages.

Although Jacobs stopped short of claiming that mixed marriages were reproductively incompatible, he nevertheless used his data to argue for the relative infertility of such unions. He concluded that Jews and gentiles were in some way too different from one another to reproduce successfully together, and that real fertility occurred only between Jews—a clear sign that they were a pure race. Jacobs was unable to attribute the relative infertility of mixed marriages to social and environmental causes, deciding in the end that it was "a phenomenon indicating racial differences."[64] It is striking that Jacobs failed to recognize the possibility that social causes such as urban living, the deliberate limiting of family size so as to improve material comfort, and increased religious laxity, especially in Western Europe, may have been responsible for this state of affairs.

Most researchers, including Jacobs, remarked on the greater proportion of boys born to Jewish parents. Among both Jews and gentiles in this period, male births exceeded female births. The Jews, however, showed even higher male birthrates than did neighboring populations.[65] The causes for the predominance of one sex over another were virtually unknown, but various scientists tried their hand at explanation. The Englishman Michael Thomas Sadler and the German Johann Daniel Hofacker, for example, independently arrived at the conclusion that it was caused by early marriage. This hypothesis was rejected by Jacobs on the grounds that early marriage was common in Russia among the Christian population, and yet they did not have as wildly disproportionate a number of male births as did the Jews. Another researcher, one J. Platter, argued that the less difference there was in the ages of the parents, the greater the likelihood of boys being born to the couple. This rather speculative piece of scientific deduction was also dismissed by Jacobs, on the basis of statistics for Budapest, where it was found that although Jewish births were predominantly male, the age difference of the parents was greater than that between non-Jewish spouses. Gustave Lagneau in France attributed the phenomenon to the observance of the Jewish ritual of *niddah*, sexual abstinence from just before to just after menstruation. E. Nagel in Budapest gave two reasons, having more to do with morality than biology: the greater care Jewish wives take of their health and the lower number of illegitimate births among Jews.[66] None of these, of course, explain the greater number of Jewish male births.

Jacobs suggested that the phenomenon was partly the result of error in the registration of female births. But the fact that the predominance of male births among the Jews was so widespread—in Eastern as well as Western Europe, cities as well as villages—led him to state that "although it is probable that the superiority is but slight, . . . its uniformity renders its existence undoubted." In the end, Jacobs felt forced to concede that excess male births "is one of the few biostatical [*sic*] phenomena which seem to be distinctively racial."[67]

Jacobs thus brought to a close what was to date the most exhaustive, wide-ranging, and scientifically objective analysis of European Jewish social life. He had shown himself to be neither a slave to current racial beliefs nor a crude biological determinist. Indeed, he found only four biostatistical points that could be attributed to racial causes: the lower number of multiple births, the relative infertility of mixed marriages, the greater longevity of the Jews, and their alleged morbidity or susceptibility to certain illnesses. Aside from these points, Jacobs maintained, almost all the data presented in his seven studies proved that peculiarly Jewish characteristics, most often attributed to racial difference, were in fact the product of the Jews' unique social circumstances. Jacobs identified the number of city dwellers among them and their relative poverty, as the two factors that left the greatest imprint on the vitality and physical characteristics of the Jews.[68] But although he was firmly grounded in a Lockean and Condillacian philosophical tradition that stressed the decisiveness of environment in shaping humankind, he maintained a belief in the concept of race, the permanency of certain racial characteristics, and the racial purity of the Jews.

Jacobs now concentrated on the second part of his scientific work: the racial characteristics of the Jews. He spent three years, from 1882 to 1885, amassing and analyzing a wealth of data from Great Britain and the Continent. On February 24, 1885, he presented the findings of his investigations to the Anthropological Institute in London. On that Tuesday evening, Sir Francis Galton, president of the Anthropological Institute, occupied the chair, and a distinguished audience of scientists and scholars was in attendance to hear two lectures on the physical anthropology of the Jews. The particular theme to be pursued that night was whether the Jews could be considered a pure race. Noting the uniqueness of the occasion, Galton opened the meeting by hinting at his own belief in the permanency of the Jewish type:

> Ladies and gentlemen, when travellers communicate to us descriptions of the races they have travelled amongst, we have commonly to regret that no members are present of the race which they describe, and again, however keenly they may have observed, their information has been based only on cursory acquaintance. On the present occasion we have no such regret. We are honoured by the presence of many eminent Jews, among whom is the Delegate Chief Rabbi, Dr. Adler, and the authors of the papers to be submitted are themselves Jews who have made the history and characteristics of their people a special study for many years. The chief topics will be whether the belief that the Jews have kept themselves free from foreign admixture for thirty-five centuries can be historically justified and whether the Jewish type of the present day has become so firmly established as to have acquired a centre of stability of its own, refusing to blend freely with

alien races whenever the artificial barriers that have kept them secluded have happened to be broken down.[69]

First, a paper by the distinguished Oxford scholar and librarian at the Bodleian, Adolf Neubauer, was read in the author's absence. On the basis of the historical evidence—from both Jewish and non-Jewish sources—suggesting widespread patterns of intermarriage, Neubauer argued against the notion of the purity of the Jewish race. In his opinion, only intermarriage and climate could explain the existence of so many different appearances among the Jews in so many countries, especially the Sephardic-Ashkenazic split.[70]

After Neubauer's short paper criticizing the concept of Jewish racial purity, it came time for the highlight of what was called the Anthropological Institute's "Jewish evening," and the podium was given over to Joseph Jacobs. Jacobs delivered a long, meticulously researched paper entitled "On the Racial Characteristics of Modern Jews," in which he argued that the Jews were racially pure and constituted a single race. Although he separated Sephardic from Ashkenazic Jews on historical and anthropometric grounds, he maintained that Jews nevertheless presented a single type whose varieties were interfertile—a crucial factor in determining purity of race.

The belief in racial purity that Jacobs espoused did not necessarily mean that he subscribed to a philosophy of biological determinism. In fact, nothing is further from the truth. The first part of his talk was essentially a recapitulation of his earlier statistical studies. His aim was to accentuate what he called secondary racial features, and here he was quite clear as to whether nature or nurture held sway: "What are the qualities, if any, that we are to regard as racially characteristic of the Jews? Much vague declamation has been spoken and written on this subject. All the moral, social, and intellectual qualities of Jews have been spoken of as being theirs by right of birth in its physical sense. Jews differ from all others in these points, it is true.... But the differences are due, in my opinion, to the combined effect of their social isolation and of their own traditions and customs, and if they have nowadays any hereditary predisposition towards certain habits and callings, these can only be regarded as secondarily racial, acquired hereditary tendencies which cannot be brought forward as proof of racial purity."[71]

The paper was divided into three sections: vital statistics, anthropometry, and historical data. Concerning vital statistics, the only new addition that Jacobs made to his previous research was a section on Jewish longevity and diseases. Jacobs rejected some of the wilder exaggerations of longer Jewish life span that had become standard in the comparative studies on Jewish-Christian mortality.[72] He cited the case of impoverished Jewish masses in Galicia and Russia, where there were greater proportions of Jewish day laborers and hence shorter life spans. The cramped, unhygienic conditions under which most Jews lived, he asserted, certainly could not promote longevity.

In Western Europe, the United States, and Australia, where Jews appeared to display longer life-spans, this again was not the result of race or any particular biological adaptation but rather was owing to "moral and social causes." In detailing those causes, Jacobs drew the kind of idealized portrait of Jewish life typical in the work of Jewish race scientists:

> Jews do not lead "dangerous" lives in the insurance sense (sailors, soldiers, firemen, miners, &c.). The trades which they do exercise, except that of tailoring, seem more long-lived. Further, the Jewish nature does not seem to require stimulants, and Jews are markedly free from alcoholism. The tranquilizing effects of Jewish family life, the joyous tone and complete rest of the Sabbath and other festivals, the unworrying character of the Jewish religion, are all important in the difficult art of keeping alive. The greater care taken of Jewish women, who more rarely take to manual labour, aids also in producing good results in the tables of mortality. I attribute much importance, too, to the strict regulation of the connubial relations current among Jews.[73]

The issue of the supposed immunity or susceptibility of Jews to certain diseases had like so much else to do with the Jews, been attributed to racial factors. Jacobs rejected the claim of some in the medical profession that Jews were naturally immune to cholera and tuberculosis, providing a wide array of figures to the contrary. He also informed the Anthropological Institute that even conditions traditionally associated with Jews, such as insanity, blindness, deaf-mutism, hemorrhoids, and diabetes, could "be traced in part to their life in towns, their mental activity, and exciting occupations."

In concluding his discussion of Jewish "vital statistics," Jacobs sounded a note of caution for his own Anglo-Jewish community as well as the many Central and West European communities he studied. He warned that "so far as Jews enjoy certain vital advantages over their neighbours these depend on the simple antique virtues and customs of the Jews and

Jewesses of past and present. These advantages will persist as long as the virtues remain, and disappear, as in some respects they are disappearing, when the bonds of religion and tradition are relaxed."[74] In other words, increasing assimilation would result in the loss of those socially induced, secondary racial characteristics that had proven advantageous in the Jewish struggle for existence.

But if the majority of Jewish behavioral characteristics was based on social and environmental causes, could it be determined that the Jews display any purely racial characteristics? For the race scientist, an answer to this question had to be based on comparative body measurements. In the late nineteenth century, Europe's conscript offices were responsible for extensive anthropometric surveys of Jews. These large-scale efforts to determine the eligibility of Jews (and others) for army service were often conducted by anthropologists and their results later published in scientific journals.[75] Jacobs, availing himself of such data, found the Jews to be the shortest (with the exception of the Magyars) and narrowest of all Europeans.[76] The substandard chest measurements, according to Jacobs, were once again caused by poor nutrition brought on by poverty, city dwelling, and lack of physical exercise.[77] As far as craniometry was concerned, Jacobs found that the Jews were predominantly brachycephalic. After collating nearly 120,000 measurements on hair, eyes, and complexion, Jacobs concluded that "though Jews are darker both in eyes and in hair than any of the other nationalities, they have about 21 per cent. blue-eyed and about 29 per cent. blond-haired." They also had about three times as many redheads as Poles, Russians, or Austrians, and about half as many as Germans.[78]

These were not just meaningless statistics accumulated for the sake of playing with numbers. On the contrary, the gleaning of statistical data was central to the methodology of race science and was the basis of its claim to be scientific. Moreover, without comparative anthropometry the question of Jewish racial purity could not have been answered.

A look at the crucial anthropological indicators of hair and eye color, as well as skull shape, shows how Jacobs was able to insist on the purity of the Jewish race while at the same time arguing that all physical variations among the Jews were caused by environmental factors. It was assumed that as a Semitic people, the Jews originally had dark hair and eyes. Semitic was, of course, a linguistic classification and not a biological one. Yet the term was widely used and abused as racially signifying those peoples who spoke a Semitic language. By the end of the nineteenth century, extensive surveys by a number of anthropologists found that as many as 25 percent of all Jews had blue eyes. To these men, the divergence from the original iris coloration of the Jews was conclusive proof of racial intermixture with non-Semitic peoples, because in places where the indigenous population tended to have light-colored eyes, so too did the Jewish inhabitants.[79]

But according to Jacobs, it was not intermixture that caused light eyes and hair to be found among Jews. Rather, they were due to environmental modification. As he expounded on the mathematical principles governing the law of averages as it applied to physical anthropology, Jacobs launched a stinging attack on the French historian Ernest Renan, who, he derisively said, had "decided [about Jewish physical variation] literally ex cathedrâ: seated in his chair at the Bibliotheque Nationale, he has observed the Jewish savants who have applied for his aid, and concluded that there are several types of Jews which are absolutely irreducible to one another." Renan's frequently repeated statement that "mon opinion est qu'il n'y a pas un type juif, mais qu'il y a des types juif" was often used by those who argued against the physical unity of the Jewish people.[80]

On the mathematical principles governing Jewish variation, Jacobs noted:

> But it will be asked, and has been asked, "How will you account for the wide divergences from the Jewish type of skull, nose, eyes, hair, &c., which are shown in the statistics on these points given above, and must indeed be a matter of common observation?". . . But the question of types is a question of averages, and you cannot so easily decide upon the non-existence of a type by pointing to a few divergences from it. An organism is not a manufactured article turned out by machinery, but may modify itself and be modified by the environment, introducing a principle of variability which causes the type to develop. An organic type therefore exists not where there is no variation, but where the variations follow the law of error, and where the modulus of variation is tolerably constant. This is in the main the case with most of the anthropological measurements I have laid before the meeting, and it follows that the variations, though they may be due to intermixture, may also be merely normal divergences from the standard.[81]

Of all the parts of the body that were subjected to anthropometric measurement, the skull commanded the most attention. Contemporary anthropological theory postulated that the skull was the most stable racial characteristic, retaining its size and shape, impervious to the influences of environmental change or social and sexual selection. Thus, changed head form was thought to result only from racial miscegenation. The belief in the permanency of head shape was shaken by the findings of the German Jewish anthropologist Franz Boas. In 1900, Boas conducted his definitive study on changing head shapes. Measuring the skulls of approximately thirty thousand immigrants to the United States, Boas confirmed that the influences of environmental change from Europe to America were able to alter the form of the head significantly. He found that Eastern European Jews' brachycephalic heads became dolichocephalic, while the dolichocephalism of Sicilians showed a marked change toward brachycephalism.[82]

Jacobs's characterization of Jewish skull shape as overwhelmingly brachycephalic, or round, was widely supported by other craniometrists. Nevertheless, for those like Jacobs who subscribed to a theory of Jewish racial purity, the present-day brachycephalism of the Jews required explanation. For if modern Jews were directly descended from the ancient Hebrews, and they were Semites, why then did they not display the same dolichocephalic characteristics as other modern-day Semites, such as Arabs? The puzzle was to explain how changed head form, without racial mixture, had taken place. There were two possible answers: either modern Jews were not purely Semitic and had intermarried on a large scale, or the ancient Hebrews were themselves brachycephalic. The famous German anthropologist Felix von Luschan maintained that the brachycephalism seen in modern Jews was the result of large-scale intermarriage between ancient Hebrews and brachycephalic Hittites. Von Luschan, who obviously rejected the idea of Jewish racial purity, added that it was from this racial crossing that Jews got their peculiar physiognomy and hooked noses. As far as the brachycephalism of the ancient Hebrews was concerned, the problem that lay before nineteenth-century craniometrists was that very few examples of ancient Jewish skulls had been unearthed. Only the Italian Jewish physician and criminologist Cesare Lombroso had hitherto measured ancient Jewish skulls. He had obtained five specimens from the catacomb of St. Calixtus in Rome, dating from about 150 CE. Lombroso stressed that the age of these skulls suggested they came from a time when barely any racial mixing could have taken place. The cranial indexes of the skulls were 80, 76.1, 78, 83.4, and 75.1—an average of 78.5. With two being brachycephalic and three mesocephalic, the argument for the predominance of dolichocephalism among ancient Jews could not, even in the face of such a small sample, be sustained. Thus, Lombroso's study confirmed that ancient Jews may have been predominantly brachycephalic.[83]

Anticipating Boas's findings by almost a decade, Jacobs dismissed the theory of head form stability, maintaining that environmental and social forces could effect cranial divergence from an original type. He had his critics. In 1898, America's premier student of race, William Z. Ripley, rejected Jacobs's contention that the Jews were racially pure. In a series of articles called "The Racial Geography of Europe," carried in a special supplement on the Jews in *Popular Science Monthly,* Ripley asserted that because of almost universal Jewish brachycephalism, "nothing is simpler than to substantiate the argument of a constant intercourse and intermixture of Jews with the Christians about them all through history, from the original exodus of the forty thousand [?] from Jerusalem after the destruction of the second temple."[84] According to Ripley, this was how Jews had become round-headed. Not only were Jacobs's arguments erroneous; his methodology was "radically defective."[85]

Jacobs shot off a blistering reply. Ripley's paper showed "a daring disregard for logic." He objected to Ripley's claim that the original, long-headed Jews had mixed so substantially with brachycephalics "that all signs of racial purity have disappeared." He also cautioned readers that his methodology was more satisfactory than Ripley's because whereas the latter came to the material only as a scientist, Jacobs had not only employed the scientific method but had also "approached the subject as a student of history." This provided him with an advantage, for the historical record showed "no evidence of any such large admixture of alien elements in the race since its dispersion from Palestine." For him, "the Jews now living are, to all intents and purposes, exclusively the direct descendants of the Diaspora." Jacobs, who early in his career had made a claim for the universal instructiveness and applicability of Jewish anthropology, sensed that this debate over Jewish racial purity was "of exceeding interest within the anthropological sphere itself. Professor Ripley assumes that round heads beget round heads, and long heads descend from long heads for all time unchanged. That appears to carry with it the assumption that no amount of brain activity can increase the mass of the brain, that skull capacity has no relation to mental capacity, and that alone among the organs of the body the brain and skull are incapable of growth, change, or development."[86]

Jacobs based his response on arguments both historical and anthropological. He reiterated what he had argued before the Anthropological Institute in 1885, that at least during the Christian era "religious antipathy has been so strong throughout that period as to form an almost insurmountable barrier to intermarriage and the consequent proselytism to Judaism which is necessary for a valid Jewish marriage." And he drew on the anthropological works of Otto Ammon (*Die natürliche Auslese beim Menschen*) and Karl Pearson (*Chances of Death*) to consolidate his argument that "in races where progress depends upon brain rather than muscle the brain-box broadens out as a natural consequence." Jacobs explained himself in purely Lamarckian terms: "If [the Jews] had been forced by persecution to become mainly blacksmiths, one would not have been surprised to find their biceps larger than those of other folk; and similarly, as they have been forced to live by the exercise of their brains, one should not be surprised to find the cubic capacity of their skulls larger than that of their neighbours."[87]

By this reasoning, Jacobs concluded that those Sephardic Jews and gentiles still displaying dolichocephalism had simply not progressed as far as those Jews and other races who had attained intellectual preeminence. In his craniometrical research on English Jewry he had found that the Sephardim of Britain displayed greater brachycephalism than the Ashkenazic population.[88] This is because the Sephardim had settled in England before the Ashkenazim and had adapted so thoroughly to British life that they had produced individuals who, by virtue of their intelligence, had achieved prominence in various fields of endeavor. For Jacobs, the march to success was accompanied by a natural, physiological metamorphosis: an increase in cranial capacity. Jacobs was responding to the then-current Aryan racial theory that vaunted the cultural achievements of longheaded Indo-Europeans as superior to those of other races. Furthermore, in the context of continued mass Eastern European Jewish immigration to Britain, and mounting British disaffection with it, Jacobs was implying that these brachycephalic newcomers would soon adapt, take advantage of the opportunities offered by English society, and make significant contributions to the cultural wealth of the nation.

In 1886, to prove his point about intellectual advancement in a propitious environment, Jacobs attempted to quantify Jewish genius. According to the statistical method devised by Francis Galton, Jacobs sought to find out "how many eminent men of certain rank exist in each million of Englishmen and of Jews."[89] Although the investigation showed that the average Jew possessed 4 percent more ability than the average Briton and 2 percent more than the average Scot, Jacobs placed little importance on the results. Nonetheless, he had great faith in the method, hoping it would lead to other comparative analyses of the intellectual abilities of nations and races.[90]

In 1900 Jacobs emigrated to New York to take up the position of revising editor of the *Jewish Encyclopaedia*. He also became head of the encyclopedia's departments of England and Anthropology. His written contribution to the enterprise included approximately 450 entries and a reader's guide. The prominent inclusion of the new field of Jewish anthropology in the encyclopedia was celebrated as a path-breaking innovation by its editors. It was also a formal recognition of the field Jacobs had initiated. As the editors wrote in the preface to the first volume:

> There remains a class of topics relating to the Jews, such as their claims to purity of race, their special aptitudes, their liability to disease, etc., which may be included under the general term anthropology. Very little research has hitherto been devoted to this subject, and it is in this ENCYCLOPAEDIA that, for the first time, the attempt is made to systematize the existing information regarding the anthropometry and vital statistics of the Jews, and to present a view of their social and economic condition.[91]

For the Anglophone world, the *Jewish Encyclopaedia* was the most comprehensive guide to the Jewish racial question available.

In New York, Jacobs revised some of his earlier anthropological research and incorporated the findings of the twenty years of anthropological study that had taken place since his first forays into the field. He continued to display enormous intellectual energy, and his capacity to produce reams of written work won the admiration of all who knew him.[92] While working on the encyclopedia he accepted a teaching appointment at the Jewish Theological Seminary, edited both the newspaper the *American Hebrew* and the *Jewish-American Year Book*, served as style editor for the Jewish Publication Society and book reviewer for the *New York Times*, and headed the Bureau of Jewish Statistics of the American Jewish Committee.[93]

He also continued to amass material for two major anthropological-historical works that he was never to complete. Fortunately, he had privately printed the outline of his proposed *The Jewish Race: A Study in National Character*. It was to be two massive tomes of sixty-seven chapters describing the "historic causes" and racial "traits" of the Jews. It would have shown the full extent of his grasp of Jewish history, culture, and anthropology. The planned work was to contain chapters on such topics as Talmud, Mishnah, Midrash, Jewish philosophy and morals, economic, political and social history, psychology, and of course physical anthropology.

The other major work Jacobs planned, an intellectual biography of western civilization entitled *The History of the European Mind*, was intended to complement his book on the Jewish race. For Jacobs did not see Jews as existing outside history; instead, their own history made them anthropologically unique and one of humanity's most valuable resources. This uniqueness, the product of inner forces and outside pressures, created the Jews' particular historical condition, allowing them to contribute the fruits of their acquired genius to the unfolding of world culture. As Jacobs so eloquently put it in an essay on antisemitism: "In the intricate warp and woof of civilization Jewish threads have been at all times constituent parts of the pattern, and to attempt to remove or unravel them would destroy the whole design. By these contributions they have earned their right to continue to work for the European culture that they have helped to develop."[94] But this forthright statement also points out one of the limitations of Jacobs's anthropology. When it came to Jews, Jacobs was thoroughly Eurocentric; although he had published many editions of Indian, Celtic, and classical folktales, almost all of his work on Jews—historical, literary, and anthropological—focused on European Jewry, especially West European Jews.

In his paper "On the Racial Characteristics of Modern Jews" Jacobs divided world Jewry into three groups: Jews both by religion and by birth—Ashkenazim, Sephardim, and Samaritans; Jews by religion but not by birth—Falashas, Karaites, Daggatouns, Beni-Israel, and Cochin; and Jews by birth but not by religion—Chuetas or Anussim, Maiminen, and G'did al Islam.[95] The contributions of Jews in the last two groups to western civilization were necessarily limited and so did not capture Jacobs's attention. These Jews still awaited their anthropologist, but they did not have long to wait: in 1895 the Russian Jewish race scientist Samuel Weissenberg began his massive studies on world Jewry, including the "exotic" communities that Jacobs had ignored.

For now, it is sufficient to reiterate that Joseph Jacobs did respond to the crises that confronted East and Central European Jewry—poverty, pogroms, and racial antisemitism—and to the large-scale resettlement in England by many of the victims, by using sociology and anthropology to attempt to resolve the Jewish problem. He thus pioneered the field of Jewish racial science and, in so doing, invented a new mode of Jewish self-defense: the scientific apologia.

In this drawing, by the Zionist illustrator and printmaker Ephraim Moses Lilien, the woman's dark features and striped robe are meant to convey her exotic origins as a Sephardic Jew. In describing such women, the American Jewish race scientist Maurice Fishberg wrote, "The traditional Semitic beauty, which in women often assumes an exquisite nobility, is generally found among these Jews, and when encountered among Jews in Eastern or Central Europe is always of this type. Indeed, it is hard to imagine a beautiful Jewess, who looks like a Jewess, presenting any other physical type."

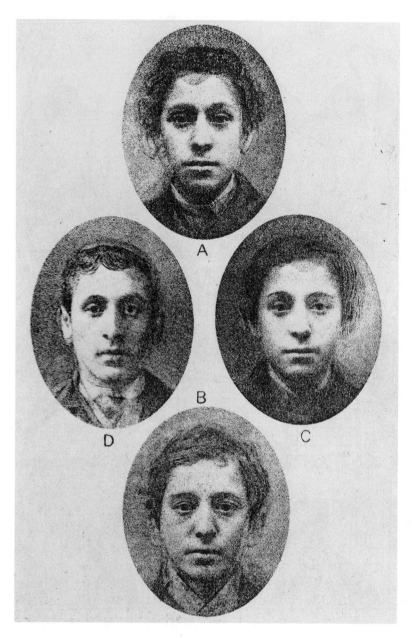

At Jacobs's request, Francis Galton took a series of photographs at the Jews' Free School in London in 1891. Galton then superimposed the ten original photographs to produce these four composite shots, which he claimed were representative Jewish types. The notion that essential Jewish features were to be found in all Jews was a central belief of race science.

Race scientists could see almost anything they wished in the objects of their gaze. Sigmund Feist chose to represent Jews who happened to have thick lips and frizzy black hair as "Negroid" and those with blond hair and long heads as "Nordic." Physical anthropologists were notorious splitters and detected countless "racial" types and subtypes, even within German Jewry.

Race science was highly subjective and prone to fanciful aesthetic judgments. The "Jewish type" was regarded by many anthropologists as prepotent and so dominant that many scientists held that some non-Jews even displayed it. The legend of the Ten Lost Tribes of Israel provided impetus for the view that certain peoples, among them the English, the American Indians, the Mayans, the Baluchis, and even the Japanese resembled Jews because they were descendants of the tribes. These two pictures appeared in Jewish race science texts with captions instructing the viewer to pay particular attention to the "Jewish physiognomy" of the subjects.

This scene from a bas-relief depicts a captive of Sennacherib's siege of the Judean city of Lachish in 701 B.C.E. Archaeological evidence such as this was often used by race scientists to compare ancient Israelites to contemporary Jews and thus demonstrate the supposed stability of the Jewish type. The use of such illustrations was also a political statement: I. M. Judt, a Zionist, sought to establish the racial affinity of contemporary Jews with the original Jewish inhabitants of the Land of Israel.

Jewish race scientists, with the aid of photography, introduced the world to many different kinds of Jews. Li King Sheng and his son Li Tsung Mai were members of the Jewish community of Kaifeng. Li Tsung Mai was circumcised at the age of twelve and given the Hebrew name Shmuel.

Race science would have placed Theodor Herzl, a Central European, in a different racial category from the two Caucasian delegates to the Sixth Zionist Congress who are at his sides, and yet the three men seem almost identical triplets, with only their dress betraying their national origins.

In books on race, the deliberate juxtaposition of photographs of different types of Jews often betrays a political agenda. Maurice Fishberg shows American Jews as clearly acculturated, secular, and solid bourgeois citizens: the woman is fashionably dressed, and the man, contrary to strict religious tradition, is clean-shaven. The Polish Jew in Jerusalem, by contrast, has a long white beard, which stands out against the background of his rumpled black outer garment. On his head is the fur cap traditionally worn by Hasidim, the *shtraimel*. Fishberg could easily have found pictures of modern Polish Jews like the American Jews he portrayed, but his aim was to express his profound faith in the United States and the process of assimilation there, and to show the inability of Zionism to transform the Polish Jew.

In this remarkable plate, a Jewish girl is posed on a photographer's set, leaning against a prop of fluted columns while standing on an Oriental rug, against a backdrop that appears to be a painting of a medieval German tower. She is a model of studied casualness. The photographer, having wrenched her from her North African environment, has sought to "tame" and "civilize" her.

FIVE

Samuel Weissenberg: Jews, Race, and Culture

"Of us three dream hunters," the Romaniot told the kaghan, "the only one you Khazars have no reason to fear is me, a rabbi; for neither a caliph, with the green sails of his fleet, nor a Greek emperor, with a cross over his armies, stands behind the Jews. Behind Constantine of Thessalonica come spears and cavalry, but behind me, a Jewish rabbi, trail prayer shawls."
Milorad Pavic, Dictionary of the Khazars

Joseph Jacobs's anthropological writings on the Jewish racial question, though broad in scope, suffused with a wealth of historical knowledge, and enlivened by wit and sensitivity, belonged to a previous era of scientific writing. The anthropology that he wrote was accessible to all, "popular" in the best sense of the word. By being published in such places as the popular Jewish press and encyclopedias, Jacobs's science retained its humanistic qualities and mass appeal. Indeed, it still bore the air of serendipity of the earliest British anthropological literature. But by the 1880s scientific writing underwent dramatic methodological change.[1] More and more, science came to be written for the guild alone. Articles were now published mainly in professional journals, and in the case of physical anthropology scientific writing became ever more quantitative and ever less popular.[2]

The pioneer of this new, truly "scientific" Jewish race science was Samuel Weissenberg. A Russian Jewish physician, Weissenberg expanded the study of the physical anthropology of the Jews, making it more quantitative and weaning it away from the western-Eurocentric view of Jewish ethnicity propounded by Jacobs. He visited scores of Jewish communities in Central Asia and the Middle East, scientifically analyzing and describing the physical characteristics of their members. He became the most widely traveled of all the Jewish anthropologists, the most familiar with the disparate elements that made up world Jewry, and the world's foremost expert on the physical anthropology of the Jews.[3] A celebrated scientist who received many academic honors and was widely consulted within his profession, Weissenberg has deservedly been called the "most distinguished of that first generation of Jewish anthropologists."[4]

Samuel Abramovich Weissenberg was born in 1867 in the Ukrainian city of Elizavetgrad, where he died in 1928. The city had been established only in 1754 and thus had a relatively short history of Jewish settlement.[5] The first immigrants arrived at the end of the eighteenth century, and in 1803 the city was home to about five hundred Jews. The nineteenth century witnessed the steady growth of the Jewish community, and by the time that Weissenberg was born Elizavetgrad had over eight thousand Jewish inhabitants.[6] This young Jewish community, without a long and vigorous religious tradition, was home to free-wheeling entrepreneurs and businessmen, not rabbis and scholars.[7] Consequently, the pull of acculturation and russification grew strong by the end of the century. At the start of his career Weissenberg was very much a child of his city, committed to the secularist path, but by the end he had become an observant Jew. This intellectual journey gave shape and direction to his anthropology, much as his coming from Eastern Europe separated him from Jacobs in England and his colleagues in Central Europe.

Although Weissenberg attended the Seventh Zionist Congress in Basel in 1905 as the Elizavetgrad representative, he never became a vocal Zionist and seemed a tepid supporter of the movement. His commitment to Diaspora Jewry was far greater. Even after undergoing his nationalist conversion, Weissenberg continued to display a sense of loving wonderment at the multifaceted grandeur of Jewish existence. He was particularly fascinated by the actors who played out the drama of Jewish life. They all appeared so different, expressed themselves in so many disparate languages, celebrated so many varying customs, and yet each had a rightful, indeed necessary and integral part to play.

But what was it that bound together this Babel of Jewish existence? Weissenberg believed that it lay in the discovery of a homogeneous essence that ran through world Jewry. The search for an essential "Jewish type" *(jüdische Typus)*, an "authentic" or "original" Jew *(Urjude)*, was to be his anthropological quest.

Like most of his fellow Jewish anthropologists (but not Jacobs), Weissenberg was a medical doctor. And like thousands of other young Russian Jews aspiring to professional careers, Weissenberg was a victim of the *numerus clausus;* his access to higher education in Russia severely limited, he was forced to seek medical training in the West.[8] In 1884 this miller's son arrived in Germany, where he attended the Technische Hochschule in Karlsruhe for three semesters, studying mathematics and natural sciences, and then Heidelberg University, at the time one of Europe's finest medical schools. Weissenberg graduated in 1890, having majored in surgery, with minors in internal medicine, psychiatry, anatomy, pathology, and physiology.[9] His dissertation, a study of dyslexia-based reading disorders, was an early contribution to the study of aphasia.[10] His willingness to expand the horizons of a very narrow medical field presaged the advances he was soon to make in another budding discipline, Jewish anthropology. After graduation Weissenberg returned to Elizavetgrad, where he practiced medicine but devoted more and more time and energy to anthropology.

Sadly, having left Russia as a teenager because of official discrimination, Weissenberg came to Germany at a time when the forces of political and racial antisemitism in the Kaiserreich were peaking. Weissenberg and other Russian Jewish students in Germany were caught in a triple bind. Not only were they exposed to general harassment as Jews, but they were also subject to stringent government regulations pertaining to East European immigrants. Yet even in this category they were treated more severely than gentile Poles, who were recruited to work in various sectors of the German economy and were thus regarded as essential. Eastern Jews, on the other hand, were primarily peddlers, merchants, religious functionaries, students, and intellectuals. Their economic utility was therefore not readily apparent.[11] In particular, Russian Jewish students were subject to a particularly virulent antisemitic offensive at German institutions of higher learning.[12] The massive influx of foreigners attending German institutions coincided with the explosion of higher learning that took place in Germany between 1870 and 1914. Between these years, enrollments at universities and technical schools mushroomed from fourteen thousand to sixty-one thousand students. The influx of foreign Jewish students exerted considerable pressure on limited educational resources. By 1912, over twenty-five hundred Russian Jews were studying at German universities and technical schools. Although they made up a fairly small portion of the total student population in Germany, Russian Jewish students were concentrated at institutions in the east of the country; two-thirds of them attended the universities at Berlin, Königsberg, and Leipzig. They also tended to enroll in the medical faculties. By 1911, for example, 85 percent of Russian Jews at Prussian universities studied medicine. At non-Prussian institutions, such as Weissenberg's alma mater, Heidelberg, the figure was as high as 90 percent. This confluence of events, plus the competition created by the presence of such Jews, created a controversy in both the academy and the government which became known euphemistically as the *Ausländerfrage*, or "foreigner question." The agitators focused on the Russian Jewish students as the primary source of their grievances. Weissenberg, who was one of these students, was therefore a direct victim of a discrimination that defied geographical borders, national affiliation, and social status.

In Germany, antisemitism was often shrouded in the supposedly objective language of science. Weissenberg, with his training in German science and medicine, could see through the obfuscations of German scientific language to discern the antisemitic motives of the writers. Indeed, he ventured to counter the claims of a hostile scientific establishment by employing the language of German race science to the advantage of his own people. Weissenberg explicitly proclaimed that the nature of his task was to do battle against German antisemitism. He declared that the professional debates over Jewish origins extended far beyond the limited discussions of the race scientists. Weissenberg lamented that "the views about the Jewish type waver between a strict affirmation of its uniformity, and the complete negation thereof. . . . The fanatical Middle Ages saw in each Jew a direct descendant of Christ's crucifiers, an actual Host desecrator, and dealt with him accordingly, martyring and burning him *ad majorem Dei gloriam*. Such a sentiment, and mode of treatment is, among the *Kulturmenschen* of the twentieth century, an abomination. And in full possession of all scientific achievements, [the non-Jew] seeks today to exclude the Jew out of state and society because the Jew is a bearer of specific, and to him, peculiar, physical and mental characteristics which prevent him from living peacefully with Aryans. This is the dogma of modern racial antisemitism."[13]

Weissenberg, like many other Russian scientists of his time, wrote his scholarly works almost exclusively in German. Germany, because of its late-nineteenth-century scientific revolution, its enormous growth as a world power, the decline of political liberalism and the concomitant rise of a virulent, xenophobic, antisemitic nationalist movement, had become the European focal point of race science. The anthropology of the Jews was a burning issue in Germany, where the Jewish problem was a matter of open, nationwide debate. By training, Weissenberg saw himself as firmly within the German scientific tradition, and he reserved his defensive salvos for those whom he deemed to be the most dangerous enemy of the Jews: the Germans.

Weissenberg's focus on non-European Jewry reflects the multinational empire in which he resided. Living in southern Russia (in present-day Ukraine), with its patchwork of neighboring Jewish ethnic groups, studying them as had no anthropologist before him, Weissenberg distinguished himself as very much a Russian. West European Jewry was of marginal intellectual concern to him, while the Jews in and near his homeland remained his prime concern until the end of his life.

His first major anthropological study concerning the Jews was the expansive "Jews of Southern Russia" (1895). A remarkable anthropological text for its time, it advocated a basic transformation in the way physical anthropology was practiced, by arguing both for broadening the scope of the investigation and for introducing fundamental changes in anthropometric methodology. Instead of focusing on one particular group, as was common, Weissenberg sought to answer larger, more fundamental questions about human growth and development by using the tools of comparative anthropometry, the measurement of the individual parts of the human body.[14] His aim was twofold: to determine the conditions that affect human growth (are they racial or environmental?) and to discover how the different races developed. Only a comparative study of different peoples could provide answers. For Weissenberg, the methodological problem with physical anthropology was that when comparative, it only addressed the physical characteristics of small sample groups of grown adults. He proposed, instead, that anthropologists conduct large-scale anthropometric surveys on newborn babies of different races and then closely follow their growth over time. This would clarify whether the environment played a decisive role in the growth process. Furthermore, he exhorted his fellow scientists to study the females of a given population in order to see if their growth rates were different from those of males, and also to observe the development of "female peculiarities" (*weibliche Eigenthümlichkeiten*) and the influence they may have on the race as a whole.

The notion that certain races were effeminate, or were in some way overly endowed with female characteristics, was widespread within both serious science and racism. Quite often it was the Jews who were singled out as womanly. The charge dates back to at least the Middle Ages, when it was claimed that Jewish men menstruated. The most infamous modern expression of the concept of Jewish effeminacy is that of the Viennese self-hating Jew, Otto Weininger. In his major work, *Geschlecht und Charakter* (*Sex and Character,* 1903), Weininger constructed a philosophy that combined pathological Jewish self-hatred with intense misogyny. He developed a general theory of the relationship between sex and character based on the proposition that every human being is in possession of both male and female characteristics. The male characteristics of logic, honor, virtue, and honesty, with the capacity for genius, were contrasted with those of the woman—dishonesty and an interest only in gaining sexual gratification, either as prostitute or, through procreation, as mother. To Weininger, the Jews were a female race. Not only hostile observers like Weininger were of the opinion that the Jews displayed predominantly female characteristics. In his book on Jewish ethnography, *Der Jüdische Stamm* (The Jewish Race, 1869), the Jewish scholar Adolf Jellinek maintained that "the study of different races shows that some have innately more male characteristics and others more female. Among the latter, who bear within them more female characteristics and at the same time represent femininity among the races, are the Jews, and a parallel between them and women is certain to convince the reader of the truth of this ethnographic thesis."[15] Weissenberg's call to study the boundaries of gender roles and their relation to race was part of a larger cultural project that took note of the instructive potential of the Jews to clarify and define the subject.

Weissenberg was aware that his proposed anthropological program, which entailed measuring hundreds of people for each study, was asking a lot of his fellow anthropologists. But, in order to achieve the desired objectives on the basis of a universally accepted anthropometric schema, he saw no alternative. According to him, the entire project revolved around three points: (1) gathering data concerning the sex, color, bodily proportion, and skull size of newborn infants of different races; (2) plotting a growth chart of a person's physical development at four specific phases of life[16]; and (3) determining the influence of climate, prosperity, and occupation on the growing organism. "The Jews of Southern Russia" was so enthusiastically received that in recognition of his contribution to physical anthropology, Weissenberg was awarded a gold medal by the Moscow Society for the Natural Sciences.

In this work, which sought to provide answers to wide-ranging questions about human development, Weissenberg used as a case study the example of the Jews who came from his native region of the Russian empire. In nineteenth- and twentieth-century physical anthropology, the Jews were seen as the instructive race par excellence. There was no other historic group of people that existed in so many different countries, displayed so many physical and cultural variations, and, though in many ways resembling the surrounding populations, remained separate from them and claimed allegiance to an extraterritorial group. Thus, they came to be regarded as indispensable to any discussion of the relative influences of nature and nurture. If Jews could remain in places for millennia and yet retain their "Jewish characteristics," then the power of environmental influences to effect change was not so great. If, on the other hand, Jews had become essentially indistinguishable from their neighbors, widely accepted theories of their physiological separateness were necessarily called into question. Different anthropologists tenaciously held these contradictory views. As had Joseph Jacobs, Weissenberg too noted the invaluable part Jews could play in the elucidation of general anthropological problems: "Anthropologically, the Jews are without a doubt one of the most interesting peoples. A people with more than a three-thousand-year history, for nearly two thousand dispersed throughout the entire world, and since then no longer a political but a closely knit religious union, absorbed by foreigners but living only under the most difficult of conditions. Separated from the rest of the world by religious laws and by hostility from the ruling peoples next to them, they offer much that is remarkable and attractive [*Merkwürdiges und Anziehendes*], not just for ethnography, but for anthropology as well."[17]

Weissenberg's study of the Jews began with the perennial question that was the basis of all studies of Jewish physical anthropology: Had the Jews retained their racial purity or not? The very question, of course, was based upon the fallacious premise that the Jews were at one time a "pure" race. Still, this question drove nineteenth-century anthropologists to reconcile the physical anomalies among the Jews by seeking their origins. What had given rise, for example, to the existence of the "blond Jews" of Europe and the "black Jews" of India and Ethiopia?

The problem of the variety of appearance among the Jews was of central concern to race science. A number of Jewish anthropologists, especially those who advocated maintaining Jewish racial purity such as Joseph Jacobs and Ignaz Zollschan, insisted upon the easy identifiability of the Jew. Jacobs believed that there was a particular countenance unique to the Jews. He went so far as to claim a Jewish "prepotency," a transmissible power that allowed "a marked and intensely Jewish cast of features and expression" to appear even in non-Jewish families "into which there has been an infusion of Jewish blood. . . . Now as reversion is mostly to the side of greater prepotency, this curious fact confirms our conclusion as to the superior prepotency of Jewish blood."[18]

Weissenberg utterly rejected this argument, which was often employed by anthropologists and others who were less than well disposed toward the Jews. He insisted that for Jews to be immediately recognizable, they would have to display a singular anthropological set of characteristics. And it was on this point that Weissenberg laid out his fundamental anthropological premise: "The Jews do not form one exact anthropological type, but are composed of several types, which are not everywhere the same."[19]

By stating this view, Weissenberg moved far away from Jacobs's theory of Jewish racial purity. But from where, and how, did all these diverse elements make their way into the Jewish people? According to Weissenberg there were three possibilities: (1) they could have been a "mixed race" (*ein Mischvolk*) from the beginning of their history; (2) through missionizing and proselytism the Jews had absorbed many foreign elements; and (3) the environment (*äussere Umstände*) had changed the Jews' original type. History may have demonstrated the veracity of the first two propositions, but it had not, to Weissenberg's satisfaction, been able to establish them as decisive. And since the third proposition was totally unproven, he determined that the task fell to anthropology to produce a vast, comparative work on the Jews of different lands in order to determine the origins of their constituent parts.

Weissenberg's study of the Jews of southern Russia was based on his examination of some 1350 Jews in Elizavetgrad, which in 1895 was home to approximately 24,000 Jews, or 40 percent of the city's population. In applying his proposed anthropometric methodology to this study, Weissenberg divided his work into nine parts, covering height, chest measurement, limbs, weight, strength, age and its relation to the development of body parts, bodily proportions of adults, influence of prosperity and occupation, and gender differences. The data on the Jews that Weissenberg gathered was then compared to figures derived from the same tests carried out on non-Jews by other anthropologists.

With regard to growth, Weissenberg demonstrated that it was not a continuous, steady process but rather was periodic, and he discerned six different periods when bodily change takes place. By comparing the measurements of southern Russian, central Russian, and Galician Jews, Weissenberg found that all three groups showed the same pattern of growth and that in bodily size they differed only slightly from one another. In attempting to discern whether there was a racial significance to the mode of human growth, Weissenberg compared the figures for Jews from southern Russia, Moscow, and Galicia with those of non-Jewish Poles, Ruthenians, Swedes, Russians, and Belgians. The ultimate height of the Jews, Poles, and Ruthenians was approximately the same, markedly shorter than that of the other groups examined. These differences, concluded Weissenberg, were governed "by external influences, and so the entire development process can be modified."[20] The determining factors included nourishment, financial status, and age at onset of puberty.

Weissenberg's study tended to show the Jews to be a relatively short European people, and he noted that, curiously enough, Jews, like Poles, Ruthenians, and Belgians, continued to grow until their thirtieth year. This observation had direct bearing on the long-standing image of the Jew as unfit for military service. The vociferous dispute about the military capability of Jews dated back to the emancipation debates in Western and Central Europe in the last quarter of the eighteenth century. Detractors accused them of chronic cowardice and a physical incapacity for warfare, and questioned their willingness to bear arms and fight on a sabbath or religious holiday and to eat nonkosher provisions.[21] Since potential Jewish conscripts were examined in Eastern Europe at about the age of twenty, they were very often found wanting.[22] Weissenberg was at pains to show that the measuring was done prematurely, and he made the point explicitly in the section on thoracic circumference.

Chest capacity was an important measurement for the military authorities in sizing up prospective recruits. The image of the weak, narrow-chested Jew, living in the unsanitary, miserable conditions of Jewish ghetto life, was widely held in both civilian and military circles.[23] The Russian revolutionary Alexander Herzen described the pitiable condition of young Jewish conscripts pressed into the czar's army: "A Jew boy, you know, is such a frail, weakly creature . . . he is not used to tramping in the mud for ten hours a day and eating biscuit . . . well, they cough and cough until they cough themselves into their graves."[24] According to contemporary anthropometry, in the well-built male the circumference of the chest should measure more than one-half of his height. Most such measurements taken of Jews in Eastern Europe found them to be well below par when compared to non-Jews. The question, however, still remained: Did this physical "deficiency" prevent the Jews from carrying out effective military service? In the armies of Russia and Austria, the authorities declared that the diminished chest circumference of the Jews did not impair their military readiness. Indeed, these two armies chose to ignore the measurement altogether, claiming that narrow-chestedness was merely a benign racial characteristic of the Jews.[25]

Weissenberg dismissed the proposition that heredity, and not environment, was responsible for the Jews' physical constitution. His comparative study showed that although Moscow Jews, in central Russia, did have underdeveloped chests, the Jews of southern Russia had chests that measured somewhat larger than the average. "Narrowness of the chest among Jews as a racial characteristic," he concluded, "appears to belong to the realm of fable.... The assertion of the absolute incapability of the Jews for military service is false." He counseled that "good military provisions and military drill would aid the development of the body far quicker" than the passage of time, which sees chest development increase with age.

It was clear, then, that it was not the influence of race but of environment—the absence of "good nourishment and clean air"—which had led to poor chest development among the Jews of central Russia and Galicia. The deficiency was quite correctable: children should be made to do lung exercises *(Lungengymnastik)* such as singing, which "exerts a favorable influence on the chest. The greater relative chest circumference of the Petersburger singers is shown by Sack [an anthropologist], and moreover, popular wisdom has composed the old rhyme: Screaming children are thriving children [*Schreikinder—Gedeihkinder*]."[26] Those who did not have the necessary conditions, and according to Weissenberg's study these were Belgians as well as Russian and Jewish schoolchildren in Moscow, paid the price with thoracic underdevelopment. Weissenberg concluded his discourse on chest measurement with an impassioned indictment: "The cause [of thoracic underdevelopment] does not lie in the race, but in the poor, yea, to judge according to the latest descriptions, terrible social and economic conditions, in the indescribable poverty and filth of the Jews in these provinces."[27]

For his study, Weissenberg also conducted strength tests. Again, these confirmed his unwavering environmentalism. Weissenberg found that the strength of Jews, white Americans, and Belgians was similar until the age of twenty, but then Jewish physical strength leveled off while the other two groups continued to gain, reaching maximum strength development at the age of thirty. What happened to Jews at the age of twenty that would impede their continued development? Weissenberg identified three possible factors. First was "the early institution of excessive intellectual work, which saw Jewish children begin school at the age of four or five." This generally took place at the "unhygienic" heder, where "the rigorousness of the teacher and the acknowledged diligence of the pupils, together with the complete disdain for all physical exercise during and after schooltime," conspired to bring about the "exhaustion" of Jewish school children. The second factor was the widespread pursuit of "such occupations which, with their lesser muscle exertion, lead to an unhealthy, sedentary way of life. I mean tailor, shoemaker, bookbinder, saddler, and similar trades, which instead of leading to the strengthening of the musculature lead to its atrophy." Not only were these trades physically stultifying, but the early age at which the poor classes of Jews gave up their studies in order to pursue them contributed to the early decline of Jewish strength. The third factor was the "early marriage of the Jews. A normal married life is without doubt an advantageous influence on the organism, yet for the Jews, especially in Russia, marriage brings about the hard struggle for existence, which, through the above causes, entirely undermines the strength."[28]

Weissenberg had dealt out a harsh social critique, squarely laying the blame for Jewish physical incapacities on the Jewish way of life—the Jewish educational system, occupational structure, and social customs. They were common complaints whose origins are to be found among the Maskilim, the exponents of the Jewish enlightenment in eighteenth-century Central Europe and nineteenth-century Eastern Europe. Their biting criticism was aimed at a complete restructuring and refashioning of not only the Jewish way of life but even the Jewish body. Weissenberg would, however, repent, and within a few years he turned his hand to Jewish ethnography. Soon he was glorifying traditional Jewish culture and campaigning for its preservation as the way to bring about a Jewish national revival.

Other themes Weissenberg investigated in his work on the Jews of southern Russia all confirmed that the physical peculiarities of Jews were due to environmental factors alone. That the influence of external conditions predominated over race was shown by the fact that marked deviations occurred within one racial group. Thus, according to Weissenberg, "the diverse forms of occupation and the various levels of prosperity probably led to colossal differences in proportion." So it was for the Jews as well. Because they were subject to analogous environmental pressures brought on by class and cultural stratification, as were non-Jews, Weissenberg could say of the Jews of southern Russia that their "development process was, on the whole, in accordance with that of the European peoples."

After analyzing his sample of southern Russian Jews according to his new anthropometric principles and comparing them to the figures garnered from tests done on non-Jews, Weissenberg was finally able to give a general description of the Jews of southern Russia: "The south Russian Jews are of medium height. In proportion to this latter aspect they have large heads and long necks, a long body, and consequently a long torso to the crown of the head. Their arms are short, their hands long, their legs short; their feet are in general the same as those of Europeans. Their chest is somewhat small, and their arm span narrow. The distance between the nipples is moderately large, and the pelvis is medium-sized. Their weight and muscular strength is less than that of most Europeans."[29]

What were Weissenberg's motives for drawing this composite picture of East European Jewry? First, general depictions of various ethnic or racial groups were very much a part of the descriptive nature of physical anthropology at the end of the nineteenth century. Such depictions, based on discrete number systems, served to differentiate one group from another and encouraged group solidarity at a time of intense nationalism by identifying and isolating the Other. Second, East European Jews formed 80 to 90 percent of world Jewry, and thus a general description of them was tantamount to a description of the average Jewish type. Weissenberg had hoped that finding the *urjüdischer Typus,* that essential element which made the Jews what they were, would perhaps unlock the mystery of the survival of the Jews as people. Third, the culture (and therefore science) of the late nineteenth century, a century marked by increasing social change and instability, was obsessed with tracing roots, establishing pedigree, and searching for ancient stability. In race science this meant searching for the *Urtypus.* This did not so much mean a quest to find an ancient paleontological representative of the whole human race as it did searching for superior and inferior races, after which the obvious contemporary political conclusions could be drawn. In this respect, Weissenberg's description of the "average" East European Jewish *Ursprungstypus* was an unexceptional and, for its day, perfectly normal anthropological exercise.

Similarly, the description must be seen in the light of what follows, for it goes to the heart of Weissenberg's own political strategy. As much as many anthropologists may have sought the shelter of objectivity through demonstrable scientific proofs, such as by anthropometry, race science was laden with prejudice and political motivation. The prevailing view that physical anthropology had of the Jews in the late nineteenth century was that they were easily identifiable because they had retained typically Jewish racial features unchanged over the course of centuries. This single "fact" was taken as proof of their racial separateness. Weissenberg challenged this point by drawing a sharp distinction between his "purely scientific" description and that provided by "travelers, [who] are rarely anthropologists and most frequently draw ethnographic portraits." For Weissenberg, only the traditional Jew, with his "religious regulations (circumcision and dietary laws), attire (side curls—pai'es), and customs, is distinguishable from his neighbor." But when the Jew is considered from the purely anthropological standpoint, it is not always so easy to separate a Jew from a non-Jew.[30] Weissenberg's observation indicates that his polemic was directed against German racial antisemitism; but it also pays attention to culturally assimilated Jews, who by appearance were indistinguishable from Germans.

By contrast, at this stage in his career he was not preoccupied with the antisemitism of Russia, which was still based on the traditional Christian image of the religious Jew and did not so readily identify Jewish difference on the basis of race. The view of Jews that the Russian statesman and jurist Konstantin Pobedonostsev expressed in a letter to Dostoevsky in 1879 is typical of the attitude prevalent in aristocratic and court circles. Complaining that the Jews were behind the social-democratic movement and were in control of the press and stock market, Pobedonostsev asserted that Jews "formulate the principles of contemporary science, which tends to dissociate itself from Christianity. And in spite of that, every time their name is mentioned, a chorus of voices is raised in favor of the Jews, supposedly in the name of civilization and tolerance, that is to say, indifference to faith."[31]

For Weissenberg, looking "typically Jewish" was not the product of a particular physiology but rather of physiognomy. The features most often identified as Jewish were outward expressions caused by the vicissitudes of Jewish historical experience:

> Centuries-long persecution and oppression, hatred from one's fellow men, exclusion from the general life and therefore a closing in on one's self, one-sided dealings in trade, and eternal fear left behind their deep traces on the entire appearance of the Jew. And they have not yet entirely disappeared, although the liberating sun of humanity has for several decades already stood in battle with the darkness and prejudice of the terrible Middle Ages. In several areas, the former has emerged victorious from this unequal battle. This, in turn, led to a Jewish response. The Jews have gradually broken

with many of their unimportant but conspicuous and constraining traditions. Unfortunately, we possess no methods to measure facial expression, and here the anthropologists must clear the field for the poets and painters. ... I have sought above to prove that mostly it is not facial construction but facial expression, and not bodily form but bodily deportment that betrays the Jew. If these change, so too does the ability to recognize disappear, and the difference [between a Jew and a non-Jew] may not be greater than that between a farmer and a scholar. I believe that everyone can easily tell the two apart, yet it is singularly certain that the defining characteristic is only a difference of the facial expression of the two. The belief in the constancy and purity of the Jewish type is a prejudice.[32]

Both modern race science and modern racial antisemitism sought to differentiate humans on the basis of physical features (often imaginary), whereas in the ancient, classical, and medieval worlds the process of self-definition, through creation of the visible Other, was executed by bestowing on the reference group outward symbols of inferiority by tattooing, head shaving, castration, or, as in the case of the Jews, special attire such as hats and badges. Nazi antisemitism of course went a step further by both identifying the physical inferiority of the Jews, and hence their essential difference from the Germans, and compelling them to wear yellow badges as outward signs of their Jewishness. The badge ordinance was, ironically, an open admission of the fundamental incorrectness and inadequacy of race theory.

Another argument that, according to Weissenberg, undermined the claims of the racial purists was that the Jews were composed of two racial types, dark, long-headed Sephardim and fair, round-headed Ashkenazim. Their obvious difference in appearance, and as Weissenberg would later demonstrate, anthropometry, was irrefutable confirmation that the Jews were a people composed of several racial types.

For Weissenberg, the several types could be differentiated from one another on the basis of hair and eye color, skull index, nose form, and extent of hair covering. He further claimed, consistent with the anthropology of his day, that there were seven different Jewish facial types.[33] Weissenberg maintained that the variegated elements that made up world Jewry were a source of pride. It was the seemingly effortless adaptability to a wide range of environments that made the Jews unique and ensured their survival.

Weissenberg concluded that the multitude of Jewish types existing among the Jews of southern Russia meant that they were not a pure race. The obvious questions that followed were, "How large was this mixing and what was its source?" and "In what relation do contemporary Jews stand to their Urtypus?" To solve these problems, Weissenberg first required a general description of the East European Jews and then one of the ancient Semitic type. Only then could the telling comparison be made. Thus even after having acknowledged the multifarious types among the Jews, Weissenberg attempted to calculate the "average Jewish type." By computing median figures for all his data he was able to come up with a composite picture, which showed that

> in its totality, all of East European Jewry appears more or less as an anthropologically uniform multitude. . . . The Jews of southern Russia (just as East Europeans generally) are, considered according to their predominant type, of medium height and brown-haired. Their head form is brachycephalic, the face oval. . . . They have a flat forehead, relatively frequently they have prominent cheekbones and a straight jaw. The direction of the eyes is horizontal, the nose is leptorrhinic [long and narrow], smaller above than below, and on the whole somewhat large and fairly prominent. Its form is predominantly straight. The lips are regular, the mouth proportionately wide, and the ears are of medium size.[34]

How did this description conform to the ancient Semitic type? As hardly any ancient Jewish skulls had ever been found in the Near East, late nineteenth-century anthropology took the Bedouins, who it believed had preserved their racial purity, as the model for the ancient Semitic type. Since East European Jews looked very different from modern-day Arabs, Weissenberg concluded that "the East European Jew has distanced himself widely from the Semitic type." How had this occurred? The answer was in "the peculiarly unique history of the Jewish people, its destiny and dispersion over the entire earth, its mighty engagement in general history, by its formation of a new, world-conquering religion, which in the beginning was not so different from the old one. All these factors were favorable for an intermixture with their neighbors, and could have led to the complete end of the original type."[35]

This problem of racial crossing between the Jews and non-Jews of Eastern Europe, and of where and how blond and Asiatic elements made their way into a group that was thought to have been purely Semitic in origin, led Weissenberg to a discussion of the origins of Jewish settlement in Russia. The traditional view held that the bulk of East European Jewry was formed out of those western Jews who had been driven east by persecution in the Middle Ages. Weissenberg, however, found it inconceivable that the millions of Yiddish-speaking Jews in modern Europe could have originated in the tiny Jewish colonies of medieval France.

Weissenberg cited the Kievan *Primary Chronicle* and other Old Russian literary-historical documents as unassailable evidence that the Jews had settled in Russia long before the eleventh century. Above all, it was the eighth-century conversion of the Khazars, a Turkic people who occupied the territory between the Volga and the Don, that proved that Jewish settlement was much older than previously thought and that the Jews had early on achieved a position of considerable influence there.[36] In arguing for the ancient rootedness and great importance of Russian Jewry, Weissenberg was making a claim that the Jews were an integral element on Russian soil. He asserted that with a mature national consciousness and exemplary tolerance, the Jewish community of the eighth century accepted and integrated thousands of Khazars into its ranks. When the Jewish Khazars were absorbed into the Kievan empire in the tenth century, the Jewish influence continued to live on after all traces of Khazar culture had vanished.[37]

The conversion of the Khazars to Judaism begs the question of when and whence Jews arrived in this region in the first place. In the alleged correspondence between Hisdai ibn Shaprut, a tenth-century Jewish physician and dignitary in the service of the Spanish rulers, to Joseph, king of the Khazars, Hisdai posed a series of questions concerning the Khazars and their relation to Judaism: "Is there a Jewish kingdom anywhere on earth? In what way did the conversion of the Khazars come about? Where does the king live and to what tribe does he belong? What is his method of procession to his place of worship?; and Does war abrogate the Sabbath?" In his reply, Joseph related that a religious debate was set up between representatives of Judaism, Christianity, and Islam, and that the Khazar king Bulan, and his highest officials were persuaded by the Jewish delegate to convert to Judaism. He reported that under King Obadiah the Khazars built synagogues and schools and learned Torah, Mishnah, and Talmud, thus accepting rabbinic Judaism.[38]

No one (in Weissenberg's time or now) seriously disputed the conversion of the Khazars to Judaism. But Weissenberg went further, presuming that Jews lived in the region of southern Russia even before the Khazar conversion. These had either come from the south, from the thriving Greek colonies on the Black Sea, or from the east, through the Caucasus.[39] Inscriptions and monuments in Crimea attested to Jewish settlement from the Byzantine empire in the early Christian centuries. There was also evidence of Jewish settlement in the Caucasus prior to the destruction of the second temple. According to Weissenberg, a change in the privileged conditions they enjoyed in Armenia led some Jews to migrate before the Christian era over the Caucasus into southern Russia. This early settlement was of central importance in the debate over the origins of Jewish types. As Weissenberg noted:

> Anthropology and history must go hand in hand to solve the dark question of the origin of the Russian Jews. . . . I am of the opinion that the causes for the changes of types to be found among southern Russian and East European Jews in general are to be sought in the migration of Jewry over the Caucasus and the Russian steppes. Perhaps already in antiquity intermixture took place, during the migration, and the Judaization of the surrounding peoples and the close contact with the pronounced short-headed Caucasians, such as the Turkic Khazars, can explain the almost complete short-headedness and frequent Mongoloid features of the Jews.[40]

Other authors, such as von Erckert, went even further than Weissenberg, suggesting that the ubiquitousness of Jews in these regions, plus the influential role they played, manifested itself racially, and that the "Jewish type" was to be met throughout the Caucasus, even among Moslems and Christians.[41]

In identifying the Caucasus as the cradle of Jewish civilization in Europe, Weissenberg had, by implication, accorded the same exalted position to that mountainous region as had the father of modern anthropology, J. F. Blumenbach, when in 1795 he spoke of the Caucasus as the racial cradle of the Europeans. Therefore, for Weissenberg, European Jews were intimately linked to Europeans by having been racially transformed by the intermixture that had taken place in the Caucasus; it was not West European but East European Jews, with their noble past and culturally vital present, who were the authentic creators and bearers of the European Jewish tradition.

Weissenberg was never a crude biological determinist, and he believed that forces infinitely more powerful than head shape were responsible for the continued existence of Jewish civilization. Not only did he posit that Jews had undergone a physical transformation in the Caucasus that now affected 90 percent of world Jewry; he also asserted that their subsequent migrations and large-scale settlement in the eastern part of the Continent meant that southern Russia—and the rest of Eastern Europe—was the cultural cradle of European Jewry. For him, this culture was responsible for preserving Jewish uniqueness and hence Jewish survival. From 1895 to 1905, Weissenberg concentrated solely on the study of East European Jewish cultural anthropology. After that time, and until the end of his life, he produced a vast amount of work on Jewish physical anthropology and ethnology.

Weissenberg's Jewish cultural anthropology was a complement to his Jewish physical anthropology. He sought to bolster his claims for the absolute centrality of East European Jewry by presenting its venerable heritage of cultural achievements. The project was undertaken at a time when increasing antisemitism fueled Weissenberg's doubts about the desirability of further assimilation, especially as it was occurring in Central and Western Europe. As the authentic bearers of the physical and cultural features of European Jewry, East European Jewry would serve as the model for a reinvigorated Jewish life where it was most needed—in the West. This distinguishes him from his predecessor Jacobs, who campaigned with great energy to help Russian Jewish refugees acculturate and become good English men and women. It would seem that Weissenberg rejected this essentially Maskilic view, preferring to have Western Jewry, if not emulate, then at the very least appreciate the richness and authenticity of East European Jewish life. In advocating this, he seems to have been in agreement with the aims of that large Buberian movement in the West which strived to introduce to alienated German Jews the living cultural traditions and deep spirituality of East European Jewry. To this end, Weissenberg became a major proponent of the pedagogic idea of establishing Jewish museums in Central Europe for the purpose of edifying a Jewish public estranged from its cultural heritage.[42] The work he produced in the area of Jewish cultural anthropology covered an astonishingly wide array of topics and placed him once again in the forefront of a new intellectual arena. The study of Jewish folklore, or *Volkskunde*, was in its incipient stages when Weissenberg made his first contributions to the field.[43]

Weissenberg's first foray into Jewish cultural anthropology clearly reflected his theories about Jewish intermixture with the peoples among whom Jews lived in southern Russia. His study of the amulets worn by the peoples of that region demonstrated that although perhaps physically different from one another, people were essentially psychologically uniform. Fundamental fears and emotions had led to the production of a similar array of artifacts in all parts of the world, from Russia to New Guinea. So deeply embedded were these impulses that they even defied class divisions. Humans, according to Weissenberg, were so thoroughly conservative that even amid the upper classes of the "highest" cultures, such as in Europe, one could find these objects employed in the same way as in Central Africa. The "struggle for existence" had led humans the world over to devise weapons for doing battle with the forces of evil. For Weissenberg, amulets belonged to the first rank of this weaponry.

Amulets were also widely used for warding off sickness and evil spirits, and were a common feature in the practice of folk medicine among East European Jews. Indeed the founder of Hasidism, Israel ben Eliezer Ba'al Shem Tov, was a faith healer and inscriber of charms; his first disciples were his patients and those who came to purchase his amulets. It troubled Weissenberg that amulets were used by both Jews and gentiles in southern Russia, for it appeared to him that what he regarded as the lower Russian culture had had a powerful but undesirable influence among the Jews.[44] It was also a clear indication of the persistence of superstition, which as a physician and man of science he disdained and no doubt saw as a sign of cultural backwardness in his people.

At the time Weissenberg wrote this essay, he adhered to the liberal German Jewish view that all Jews were part of a *Glaubensgenossenschaft* (community of faith) but not a *Nationalgenossenschaft* (nationality). The ghetto novelist Karl Emil Franzos articulated this stance admirably: "We are no longer a uniform people [*Volk*]. . . . The Jew who lives in the civilized countries is today a German, a Frenchman, of the Jewish faith, and thank God that this is so! We are now only a community of faith dispersed in all the lands of the World! We feel Jewish only through our faith, not our nationality."[45] But by 1900, Weissenberg came to temper this view. He became a *Nationaljude,* recognizing and celebrating the unique folkways of the Jewish people. He saw their different customs and traditions as scenes in a great tapestry: each individually was enhanced by being placed next to the others, yet in ensemble they created a magnificent and unified impression of the whole of Jewry. This is not what Franzos had in mind, for there was no possibility in his conception of Jewish life that East European Jews, for example, could enrich the lot of German Jewry. Weissenberg felt just the opposite, and his anthropology from the turn of the century was dedicated to driving home this point.

Weissenberg's subsequent work on Jewish cultural anthropology was profoundly influenced by the twin pillars of Zionism and antisemitism, and he apportioned responsibility for the belated awakening of Jewish national consciousness to both. He waxed rhapsodic over the reality of recent Jewish self-assertion:

After a long, sweet sleep, the Jewish spirit finally awakened. To be sure, not independently, but after a fairly harsh shaking on the part of [the Jews'] eternal, living adversaries. The agitation against the Jews, which, like a storm wind, seizes and pulls down everything (also that which appeared unimpeachable), has its good side as well. It sweeps away the long-held hope of assimilation and shows that the Jews, despite their best intentions, cannot perish. One may think what one will about Zionism, but it cannot be denied that it is generated by latent power and energy, and an awakened self-consciousness.[46]

Weissenberg took the opportunity to affirm his conviction that the furtherance of Jewish national rebirth depended on the energies of East European Jews, who, with their authentic *Volksleben*, formed the model of a proud and confident people. While seemingly caught up in the rising tide of Zionist euphoria, Weissenberg, as he would do in his later studies of Jewish anthropology, both cultural and physical, regarded the Zionist movement as an effective means to an end, but not an end in itself. For him, Zionism was a catalyst in the rise of Jewish national feeling, the drive to become *Nationaljuden*, not a call to terminate Jewish existence on European soil. As he maintained, with characteristic cosmopolitanism, "One can be a Jew and at the same time be a loyal German citizen, just as one can be a German and at the same time be a Swiss or Russian citizen."

In the wake of the rabid nationalism engulfing the Continent, the Jews, according to Weissenberg, found themselves not only isolated and abandoned, but also the objects of obloquy and persecution. "Under these conditions," he wrote, "the Jews finally braced themselves and declared themselves a nation." The consequences were nothing short of revolutionary, resulting in "a blossoming of Jewish scholarship and the Jewish spirit. The ancient Hebrew language is, in Palestine and Russia, no longer only diligently studied, but is used as colloquial speech. One is no longer ashamed to be a Jew. [One also] mourns that a large part of the thousand-year-old national treasures have been irrevocably lost and seeks hastily to save the remains."[47]

Weissenberg lamented that despite the establishment of new institutions dedicated to the study of Jewish folklore in such centers as Vienna and Hamburg, people still doubted whether a true Jewish folklore existed. This, he charged, was a problem of the distorted perspective held by those who lived in Central Europe: "There are no Jews, but only Germans of the Jewish faith, so there can be no Jewish folklore, just as there can be no Catholic folklore. That is the reasoning of many non-Jews, and Jews as well. But one forgets that aside from the few men of commerce, law, and medicine, there is a large Jewish mass which adheres to its traditions and ideals. They possess a life of emotion [*Gemütsleben*], and these traditions and ideals are rich, differing fundamentally from those of the surroundings, and are still to be researched."[48]

Weissenberg took seriously this claim about the importance of research in Jewish ethnography. Some of his earliest publications included a study of Yiddish folk songs from different regions of southern Russia, as well as an edition of some three hundred Yiddish proverbs.[49] By publishing this material, Weissenberg followed directly on the heels of the influential Yiddish folklorist Ignatz Bernstein, who in 1888–1889 published 2,056 Yiddish proverbs in the annual journal *Hoyzfraynd*.[50] Weissenberg's modest contribution to the field of proverb collecting served to reinforce his view of the separate and "authentic" development, especially linguistic, of Jewish civilization in Eastern Europe. This stood in stark contrast to the West, where—as far as Weissenberg and many other East European Jews, as well as urban West European Jews, were concerned—there were no longer any living Jewish languages. However, this was clearly not the case. In Germany, what was called Western Yiddish survived, especially in rural areas, into the twentieth century. Similarly, the unique and rich Yiddish linguistic traditions associated with Alsacian Jewry only came to an end with the destruction of that community during the Holocaust. Although he studied in Baden, where some people still spoke Western Yiddish, Weissenberg was mostly confined to the university at Heidelberg, with its *Hochdeutsch*-speaking community. Consequently he, like most other researchers in Jewish folklore, ignored the rich resources of what was an "authentic Jewish culture" in the middle and west of the Continent.[51]

If Yiddish folk songs and proverbs were, by their content and mode of expression, identifiably Jewish, the same unfortunately could not be said for art created by Jews. Weissenberg pondered the problem of defining the nature of Jewish art as well as setting out a program for salvaging its treasures. In his eyes, traditional Jewish art was religious in nature. Whether it was biblical poetry, the sayings of the prophets, or the later creation of material objects for the synagogue, Jewish art found its inspiration in the religious life of the people.

But for Weissenberg, the problem of Jewish art centered on the absence of a uniform style. Owing to periodic expulsions and constant contact with different populations, Jewish artists, in more recent times, often borrowed foreign styles without necessarily transforming them into works of Jewish national art. This was a direct consequence of the "drive for assimilation" (*Assimilierungswahn*) which had taken hold of the Jews, and it had not only fostered the creation of this "un-Jewish" art but had also led many Jews to declare those ritual objects they owned useless, fit only to be discarded. But happily, attitudes were now changing because "the Jews begin to once again be Jews. One speaks of Jewish art, one establishes Jewish museums. We are experiencing a Jewish renaissance. For my part, I see in the Jewish museum the mainstay of the modern Jewish movement, for a true renaissance is not possible without thorough study of the entire past way of life."[52]

Yet a Jewish renaissance from the artistic point of view did not necessarily mean a return to the singular production of ritual objects. To Weissenberg, the awakening of Jewish national consciousness was a modernist movement, and in its train the modern in art "need not be discarded, provided that it bore a specifically Jewish stamp in terms of its style or concept." What was required was a painstaking examination of the material in order to determine, among other things, the exact origin and date of completion of a particular object. This, in turn, would enable one to "separate the husk from the kernel and thus lay the foundation stones for the construction of a Jewish style."[53] The idea here was to weed out the non-Jewish influences that had crept into the consciousness of Jewish artists, eventually replacing the Jewish content of their art altogether.

The task that Weissenberg set Jewish art historians and critics was to seek out the authentically Jewish and re-create from this legacy a new Jewish style worthy of the Jewish renaissance, one that would ennoble the Jews and make them conscious bearers of the national idea. It is here that we can most clearly see the connection between Weissenberg's physical and cultural anthropology. For in both, as he had begun to do in his study of the Jews of southern Russia and would continue to do in his work on non-European Jews, Weissenberg sought to identify the *Urjude*, be it in the form of an *urjüdischer Typus*, in the case of his race science, or an *urjüdischer Stil*, in the case of his cultural anthropology. The search for roots and authenticity was a central feature of late-nineteenth-century culture, and Jews, no less than their opponents, sought out the authentic and genuine.[54]

Weissenberg continued to publish prolifically on all manner of Jewish topics. Working at a feverish pace, he wrote a string of articles touching on traditional customs associated with the birth of a child, marriage, care for the sick, and burial of the dead, the celebration of festivals and fasts, as well as the place accorded Palestine in the life and belief of contemporary European Jews.[55] Yet despite celebrating the rich national folkways of the Jews of Eastern Europe, he was not so naive as to paint life among the Jews of Russia as a picture of uncorrupted cultural purity. Even though he continued to regard East European Jewish culture as the only viable model for a rebirth of Jewish national life, he protested bitterly that even here, in this bastion of Jewish tradition, the same forces of assimilation that he had identified as having decimated West European Jewry were at work. His general view was encapsulated in a discussion of the way the festival of Purim was currently being marked:

> The persecutions in Russia have brought the joyous life of the Jews to an end. The anxiety over daily bread which is much greater than before, and the uncertainty of life in general have struck such deep roots in the soul of the Jewish people, that every bit of cheerfulness chokes on bleak reality. The earlier, merry, and beautiful festival of Purim is hardly celebrated anymore, and it is no longer remembered that at one time, life was so active on this day on the street and at home. One sees no gift-givers, no *gragers*, no Purim players. Among the rich, modern cakes replace the artistic pastries of grandmother. The modern schools seek to "exterminate every folkloristic seed." The old craftsmen who carved the noisy *gragers* have long since died. The simple Purim play no longer amuses. The little Hebrew that was in them is mostly no longer understood, and the public seeks naturalistic attractions. So two powerful enemies are working to destroy the Jewish people: the pseudo-culture of the upper classes and Haman's repeated rebirth among the lower peoples.[56]

The twin experiences of antisemitism and assimilation had deeply affected Weissenberg's work and led to a change in emphasis in his anthropology. After initially practicing a science that addressed broad, universal themes—his first forays into physical anthropology did not address Jews and were concerned with methodological issues—Weissenberg came to practice one that was far more insular and more singularly focused on the Jewish people, especially their cultural achievements in Eastern Europe.

After the publication of these works on the cultural anthropology of the Jews, Weissenberg returned to their physical anthropology. This aspect of his work also reflected his increasingly narrowed focus. In his race science from this point on, no less than his ethnography, he also looked to sort the wheat from the chaff by identifying the *urjüdischer Typus* through a thoroughgoing comparative investigation of European and non-European Jewry.[57]

In 1908 Weissenberg's greatest contribution to the study of the physical anthropology of the Jews began in earnest. That year he received a stipend from the Rudolf Virchow Foundation, which enabled him to travel to Turkey, Syria, Egypt, and Palestine, where he undertook anthropometric studies on 690 individuals. Of this group, 561 were Jews, 45 were either Samaritans or Karaites, 64 were Arabs, and 20 were Armenian.[58] Weissenberg's travels and his quest to find the Jewish *Urtypus* were based upon his assumption that the Sephardim, one branch of the descendants of the ancient Israelites, had maintained a certain level of racial purity and uniformity. These Jews would most likely be found either in or within relatively close proximity to Palestine, and it is to those places that Weissenberg went.

Like nearly all other race scientists who concerned themselves with the physical anthropology of the Jews, Weissenberg's main interest lay in the origins of the two distinct Jewish types, Sephardim and Ashkenazim.[59] Although Weissenberg had long since concurred with that school that maintained that the two groups constituted two separate anthropological types, the central question was when and how the division took place. The two competing explanations were those of Weissenberg and the distinguished German anthropologist Felix von Luschan. Weissenberg maintained that the ancient Israelites were dolichocephalic, and that it was the Sephardim who most closely resembled them, whereas the Ashkenazim had deviated from this original type by intermixing with the brachycephalic peoples of the Caucasus. By contrast, von Luschan expressed the widely held opinion that the ancient Hebrews had never been a pure dolichocephalic race but were, rather, an amalgam of at least three different races. Consequently, according to von Luschan, "the modern Jews are composed, firstly, of Aryan Amorites, secondly, of genuine Semites, and, thirdly and primarily, of the descendants of the Hittites. Along with these three most important elements of Jewry, other admixtures have taken place during the course of their millennial diaspora."[60]

Even though von Luschan argued for the relative stability of the Jewish racial type since the crossing had occurred in prehistoric times, Weissenberg conducted a vehement campaign against this thesis in scores of articles. But he too had once denied the racial purity of the East European Jews. Employing an argument of racial mixing identical to his rival's, he suggested that since their crossing with Caucasians had probably taken place before the destruction of the first temple in 586 BCE, and with little crossing afterwards, the Jewish type in Europe was to be regarded as stable and relatively uniform. What was it, then, about the arguments that created the intellectual chasm between them? The two scientists were separated by a single point—Weissenberg's insistence that the original Jews had been a pure race. To gain support for his thesis, he set out for Palestine to find and study their descendants.[61] In Palestine, too, he would encounter other autochthonous groups with whom to make anthropometric comparison.[62]

Visiting Jewish communities such as Peki'in, Safed, Shefa'amr, and other scattered groups in the Galilee, many of whom had remained in the land since the destruction of the second temple, Weissenberg was sure that he had discovered perfect examples of what he called *Judaeus primigenius*.[63] Essentially indistinguishable in language, dress, and customs from the neighboring Arabs (whom he assumed must resemble the original Canaanites with whom the Hebrews commingled upon their entry into the land), Weissenberg asserted that these Jews differed markedly in appearance from their brothers and sisters in Europe.

To test and prove his hypothesis concerning the Jewish *Urtypus*, Weissenberg undertook a comparative anthropometric analysis of Jews, Samaritans, and Arabs.[64] All three groups were dolichocephalic, of dark complexion, with no blondes, and of medium height; all displayed, in different proportions, the straight "Semitic" nose. Thus, on the basis of his own measurements, Weissenberg was able to declare confidently that "our analysis proves that one could give up on von Luschan's Hittite theory."[65]

While in Jerusalem and Jaffa, Weissenberg measured fifty Yemenite men and fourteen women. The Jewish community of the southern Arabian Peninsula was of particular anthropological interest because of its great antiquity and relative isolation. By the early twentieth century, a considerable literature on the community had already developed, but anthropologists were divided as to the racial identity of the Jewish population.[66] The two main schools of thought held that either the Jews of Yemen were a distinct race unto themselves, differing from the other major Jewish racial types, or they were racially Arabs who had long ago converted to Judaism.[67] For the time being, Weissenberg inclined toward the latter view but remained quite unsure. By comparing Yemenite with southern Russian Jews, he found that the former were of short stature, displayed pronounced dolichocephalism, had dark complexions, small oval faces, and small "Semitic" noses. This was a markedly different picture from the one he drew of European Jews, and after closely comparing the two populations he pronounced that "the above analysis allows one to conclude without hesitation that between the European Jews and their Yemenite coreligionists no physical relationship exists."[68]

The differences Weissenberg detected did not resolve for him the issue of Yemenite Jewish racial purity. Within months, however, all hesitancy on his part about the purity of the Yemenite Jews had disappeared as a result of his study of the two main groups of Caucasian Jews, the Grusians and the Mountain Jews. So thoroughly different in appearance were the Caucasian and southern Arabian "branches" of the Jewish people that he became convinced that it was the isolated Yemenites who had maintained their racial purity, in contrast to the Jews of the Caucasus, who had been modified and transformed through their encounter with the hyperbrachycephalic Caucasians.[69] This conclusion, of course, was once again predicated on his rejection of von Luschan's theory of proto-Hebraic brachycephalism. So convinced was Weissenberg of the Semitic character of the original Jewish type that he even suggested that the Jews of the Caucasus were merely "Judaized Caucasians." The indirect proof of this possibility was the early appearance of Christianity in the region, for as Weissenberg asserted, Christianity only arose in those areas that had already been Judaized and had their soil prepared by Jews for the planting of Christ's teachings.

Before the Caucasian trip, Weissenberg, while still in Jerusalem, made for the Bukharan quarter of the city. There he measured eighteen Jews from Central Asia. Coming from Bukhara, Samarkand, Merv, and Herat, these were the first Central Asian Jews to have ever felt an anthropometrist's calipers. Although they came from places separated by great distances, these Jews spoke a common dialect of Persian, which for Weissenberg meant that Persia was the *Stammland,* the original seat of these people. Since anthropology regarded Persian Jewry as having remained truer to the original racial type, and since Persia was regarded as the point of departure for the great migrations of Jews to Central Asia, it was of paramount interest for Weissenberg to see what physical relations existed between the *Stammeltern* and their possible *Stammkinder.*

According to Weissenberg, the only physical features that the three groups of Jews from Central Asia had in common were their medium height and dark complexion; they showed wide differences in all other indices. He concluded that "among Central Asian Jews we do not have a uniform, but a mixed group. This is expressed primarily in the form of the head and face."[70] Curiously enough, noticeable similarities in the crucial skull index and the nasal index linked Central Asian and southern

Russian Jews. This led Weissenberg to claim that Persian Jews, during their migrations in Central Asia, mixed with the indigenous brachycephalic populations.[71] This was, of course, a recapitulation of his basic theory. For him, the fact that some Persians had retained their narrower head form indicated that among Central Asian Jews, this "transformation of the originally long, Jewish head into its current broad form" had taken place as a result of intermixture. This theory was made plausible by the fact that the "present population of Central Asia, with its pronounced brachycephalism, belongs to the Mongolian race."[72] Weissenberg later confirmed this finding by comparing Persian Jewry with the Jews of Central Asia and the Mountain Jews of the Caucasus and discovering "noticeable differences in the structure of the head. The Persians were mesocephalic, the Central Asians brachycephalic, and the Mountain Jews almost all hyperbrachycephalic. These differences can hardly be explained other than through intermixture with broad-headed elements."[73]

Thus far, Weissenberg's results showed that only the native Jews of Palestine, and perhaps those of Yemen, corresponded to the idealized Jewish *Urtypus*. The physical appearance of those Jews in western and northern Persia and Central Asia seemed to confirm his theory of the Jewish racial split. What was left for him to do was to examine that group which nearly all race scientists identified as most closely resembling the original Jewish type, the Sephardim. As he had stated some years before, anthropological research had mainly focused on Ashkenazic Jewry, so he took as the primary task of his Middle Eastern tour the gathering of anthropological data on the Sephardim and the elucidation of the problematic anthropological split between them and the Ashkenazim.

Weissenberg measured 175 Sephardic Jews from Constantinople and Jerusalem. So physically similar were the two groups that Weissenberg was prepared to regard them as "forming a uniform type." He described this group as being of medium height, mesocephalic, commonly having the "Semitic" nose, and brunet. After comparing the results again with his southern Russian sample, Weissenberg declared that "as the older authors have asserted, the Sephardim have preserved themselves more purely to the Semitic type than the East European Jews."[74] Thus, on the basis of anthropometric measurement, Weissenberg verified the anthropological split between Sephardim and Ashkenazim and established that the former most closely corresponded to the original Jewish type, thus contradicting the theory of his rival, Felix von Luschan. But in reality we can see just how tenuous Weissenberg's claims were. His descriptions of Sephardim, with their supposedly greater racial purity, hardly differ from his descriptions of Central Asian, Yemenite, or Mesopotamian Jews, whom he classified as either being mixed—possibly not Jews in the racial sense at all—or even descendants of the Ten Lost Tribes.[75]

A further example of the methodological shakiness of the race scientist's enterprise appears in Weissenberg's work on Syrian Jewry. Given the geographical proximity of Syria to Palestine, Weissenberg assumed that it was the land to which Jews probably first went upon being driven from their ancestral homeland.[76] His anthropometric study of these Jews, which was preceded by a brief historical outline of Jewish settlement in that country from biblical times to the seventeenth century, which emphasized the large influx of Jews from Spain after the expulsion of 1492, led Weissenberg to conclude that "Syria proper was never absent of Jews." Thus, ancient and continuous Jewish residence there meant that the chances of finding the Jewish *Urtypus* were extremely good.

Having established the settlement history of Syrian Jewry, Weissenberg gave the anthropometric results of the thirty Damascene and twenty Aleppo Jews he examined and measured. He found them to be totally different from each other physically. The Jews from Damascus, with a skull index of 80, were mesocephalic, while those from Aleppo had a skull index of 84.3 and were almost hyperbrachycephalic. For Weissenberg, this meant that "the Damascene Jews are less mixed, and can be considered to be closer to the *Urtypus*, while those from Aleppo have absorbed much foreign blood, especially through the immigration of the Spanish Jews."

There is much that is wrong with this reckoning. First, it is based on the immigration to Syria of an unstated and probably unknown number of individuals, whose genetic predominance Weissenberg took to have been so great as to transform the indigenous, "pure" dolichocephalic Jews in only one city into a hyperbrachycephalic population. Second, Weissenberg says nothing of the city itself, and especially of the history of Spanish Jewish immigration to it. Third, Weissenberg's theory of Sephardic mesocephalism is based upon his conviction that significant intermixture had taken place in the Middle Ages between Spanish and East European Jews.[77] The point here is obviously not to test these theories by modern science, but to point out that Jewish race science was at least as susceptible to bluster and specious argumentation as its non-Jewish counterpart. And it was most often in the search for imaginary *Urtypen* that race scientists of all backgrounds were led wildly astray.

In fact, Weissenberg's descriptions reveal very little that separates the Sephardim from the Ashkenazim: "The comparison chart teaches us that Sephardim are almost the same height and possess the same reach [as

Ashkenazim]. They have a somewhat longer and yet narrower head, which is caused by the smaller skull index. Aside from this they have the same face length and narrower cheeks. The noses of the two groups are almost of the same dimensions. The skull index of the Sephardim is mesocephalic, while that of the southern Russian Jews is brachycephalic."[78]

It is clear that even Weissenberg had difficulty in painting distinctive pictures of the two groups. To be sure, he did notice an appreciable difference in the complexions of the two, as 79.2 of Sephardim but only 58 percent of Ashkenazim were brunet. But the all-important skull index, which served to classify Ashkenazim as brachycephalic and Sephardim as either mesocephalic or dolichocephalic, were in real numbers quite close: 78.1 mm for Sephardim and 82.5 mm for Ashkenazim. So Weissenberg, like other Jewish and non-Jewish race scientists, made exaggerated claims about racial difference and relative purity, primarily on the basis of a 4.4 mm differential in skull index. For Jewish anthropologists like Samuel Weissenberg, who eagerly employed the European scientific methodology of the day, succumbing to the enticing vagaries of race science was both part of the Jewish drive for professional and universal acceptance and, paradoxically, part of the process of cultural distancing that Jews were beginning to engage in with the rise of their own national movement.

A detailed study of North African Jews in 1912, with a long introduction drawing on the most recent historical, archaeological, economic, and literary research, not only displayed the breadth of Weissenberg's learning but also proved that North African Jews conformed closely, as had the Jews of Palestine and Yemen, to the Jewish *Urtypus*.[79] The dolichocephalic Jews of North Africa were very different in appearance from their brethren in Central and Eastern Europe, and though perhaps not as "pure" as the Yemenite Jews "because they had not remained free of foreign blood," had nevertheless "experienced no or only few changes of their type because of mixture with related Hamitic or purely Semitic elements." And by contrast, one of Weissenberg's last studies on the physical anthropology of the Jews, a comparison of them with Armenians, showed the significant intermixture that had taken place between dolichocephalic *Urjuden* and brachycephalic Armenians.

Modern race science had identified contemporary Armenians as the descendants of the ancient Hittites, a finding with which Weissenberg agreed. But discussion of the details of the encounter between Jews and Armenians, which led to a loss of those authentic Semitic characteristics of the Jews, once again brought Weissenberg into heated conflict with von Luschan over the chronology of miscegenation. As Weissenberg wrote,

"The foundation of the similarity between Armenians and Jews is thus to be sought in the actual blending of both peoples. But this did not take place in the prehistoric period on the soil of Palestine, but rather in the historic period on the soil of Armenia."[80]

Weissenberg's 1915 study of the Armenians was really his last that was conducted on the methodological principles he had outlined some twenty years earlier in his study of the Jews of southern Russia. The old-style race science, based on description, comparison, and large-scale measurement, was beginning to be augmented by eugenics, with its prophylactic proposals for race preservation.[81] Weissenberg's work from this point on reflected the new trends in race science, his own medical specialization as a gynecologist, and many of his older beliefs. In a series of articles on the forms of sexual intercourse and birthrates among Jews, the sexual ethics of Judaism, and the social hygiene of the Jewish people,[82] Weissenberg combined the findings of his extensive physical anthropology, ethnography, and the new eugenics.[83]

Weissenberg was socially very conservative, and in his later years it appears he became an extremely devout man. A vocal critic of coeducation, he opposed too much social intercourse between the sexes and what he called the insidious features of the women's movement, which he felt had manifested themselves in the increase of "premature" births and even the rising number of abortions.[84] The last works of his life, before he succumbed to stomach cancer in 1928, provide a picture of a scientist-scholar with a deep knowledge of and reverence for traditional Jewish sources and the Orthodox way of life. He took pride in East European Jewry's "high sense of family," the extraordinarily low rate of illegitimate births, the minimal amount of sexually transmitted diseases, the low infant mortality rate, the moderate alcohol intake, and the apparent immunity to many infectious diseases, but asserted that none of this was due to any particular racial qualities. These features were not so apparent in the less traditional Jewish communities of Western Europe. This was proof, for Weissenberg, that any biological benefits the Jews might have enjoyed were solely founded in their religion. There had not been a "physical selection of certain racial elements, but a pure intellectual selection of those elements which see in the maintenance of Judaism a cultural entity [Kulturfaktor] and cultural power [Kulturmacht] as its life goal. "Consequently," he confessed, "for me, Judaism is a cultural phenomenon [Kulturerscheinung]."[85] For Weissenberg, all the Jewish festivals and ritual practices were grounded in reason and promoted good health.[86] It was stringent adherence to them, rather than possession of any racial peculiarities, that would ensure for the Jews a eugenically prosperous future, for "so long as the Jews not only preserve their culture, but also promote it and adapt to new achievements, they will remain ineradicable."[87]

By the end of his career Weissenberg had fully developed and defended the ideas he had maintained all along—that Judaism was a society based upon law and custom, and not one predicated upon skull size and brain weight. Of course, race science was not without importance for him either. It enabled him, or so he believed, to come to a deeper understanding of the complexity of the Jewish social organism. And despite the relegation of race science to the realm of dead (and perhaps shameful) science, in its day Weissenberg's work was on the cutting edge of physical anthropology, and he was recognized as the world's foremost authority on the Jews as a race. His wide-ranging travels brought him in touch with more Jewish communities and types of Jews than perhaps any of his contemporaries. Consequently his descriptions, particularly of the customs and culture of these communities, which have ceased to exist, are still of value to this day. He believed that religious culture had not only been the salvation of the Jews in times of antisemitic aggression but would also be the cement that would bond them together in the face of the disintegrating power of assimilation.

To Weissenberg, the strength of the Jewish people lay in their dispersion. Of all the ancient civilizations, only the Jews remained a potent force. Those who had strong territorial bonds—the Greeks, for example—had disappeared from history. The Jews, on the other hand, were in the midst of a national renaissance. But this rebirth did not require that an end be brought to that which Weissenberg believed had imparted to the Jewish people its vitality: the refraction of Judaism through the prism of Diaspora reality.

Though he had offered mild support to those who called for an end to the Diaspora—the Zionists—he was a skeptical and somewhat bemused participant in the enterprise. He even described as a "psychological riddle . . . the childishly naive, pure love of the Holy Land that still to this day, after nearly two thousand years of exile and dispersion in all parts of the world, animates every Jew."[88] He did not fully share the Zionist hope for the future because he was too mired in the glory of the Jewish past. The Zionist image of the Jewish people and their history was the opposite of the one drawn by Weissenberg and was predicated upon belief in the atrophying effects of Diaspora existence. Consequently, the Zionists' wish was to reinvigorate the Jewish *Volkskörper* by restoring it to its ancient homeland. This Zionist critique of Jewish existence also involved a biological interpretation of that past and of the present. In Central Europe, German Zionist physicians took an active part in the medical debates over the "racial" properties of the Jews. As we will see, the places in which they lived and worked and the social problems they personally witnessed within their own Jewish communities largely determined their scientific agendas.

✼ SIX ✼

ZIONISM AND RACIAL ANTHROPOLOGY

> Who you are? Perchance the son of the nervous Jewish peddler Nathan and lazy Sarah, whom he had fortuitously slept with because she had brought enough money into their marriage? No! Judah Maccabee was your father and Queen Esther your mother. From you, and you alone, the chain goes back—across defective links, to be sure—to Saul and David and Moses. They are present in every one of you. They have been there all the time, and tomorrow their spirit could be revived.
> *Theodor Lessing,* Der jüdische Selbsthass

In 1904, a little book appeared with the intriguing title, *Anthropologie und Zionismus* (Anthropology and Zionism). Written by Aron Sandler, a physician, Zionist, and later a leading member of the Jüdische Volkspartei, it was a plea for the Zionist cause, employing scientific evidence to prove that if not in a physical sense, then in a psychological one, the Jews, like all great races, possessed a unique individuality. According to Sandler, although the Jews had a particular genius, proven by their having given so much to humanity, they could soar to still greater heights. Naturally, this could only happen in a Jewish national homeland. Despite his own partisan testimony, couched here in racial imagery and language, Sandler warned against using the idea of race to foster a dangerous jingoism. He cautioned that "in every case, one kneels to the power of the facts and wants to see what really is! But one must also be careful of the antithetical evil—racial chauvinism, which today spreads itself wide [and wherein the proponents of such ideology] claim for themselves the title of noble race par excellence. It is a hideous growth on the tree of anthropological knowledge."[1]

A decade later, and after a sizable corpus of literature on the racial qualities of the Jews had been amassed, the German socialist leader Karl Kautsky published *Rasse und Judentum,* a refutation of all racially deterministic theories of Jewish behavior. Pointing out the dangers of what Sandler had called racial chauvinism, Kautsky accused the Zionists of employing race theory for their own nationalistic ends. He wrote that "characteristically enough, there is rising within Judaism, as a reaction against anti-Semitism, a similar tendency to accept and utilize the theory of race. It is a natural application of the principle: If this theory permits Christian-Teutonic patriots to declare themselves demi-gods, why should Zionist patriots not use it in order to stamp the people chosen by God as a chosen race of nature, a noble race that must be carefully guarded from any deterioration and contamination by foreign elements?"[2] Neither of these competing claims, that Zionism was a legitimate expression of Jewish national aspirations and that it was a misbegotten racial response to antisemitism, does justice to the complex discourse of race engaged in by Zionist physicians. Although the scientists fulfilled Sandler's ideal of avoiding chauvinism, they made more use of race science than as a mere tool in their project of nation building. And contrary to Kautsky's claim, Zionist race science, while sharing the methodological approach of contemporary physical anthropology, was not an imitation of the racist nationalism of his contemporary Europe.

The Background

German scientific preeminence, the high participation of Jews in the German medical profession, and the prominence in Germany of discussions about race are some of the reasons that the most vocal and influential Jewish race scientists were German Zionists. But other factors also helped make Germany a center of Zionist anthropology. It was in the center of Europe, in German-speaking cities like Berlin, Leipzig, Frankfurt, Munich, and Vienna, that the twin forces of assimilation and organized antisemitism, so evident in Britain and Russia respectively, coincided. At a time when the flourishing German Jewish community was the focus of organized political opposition and popular ostracism, Zionist physicians played a crucial role in countering what they saw as the Janus-faced enemy.

For the Zionists, Jewish life in Germany (and eventually the entire Diaspora) was threatened not only by the hostile mob but also by the prospects of "race suicide," to use a contemporary term. As far as the Zionists (and of course the religious) were concerned, German Jewry would self-immolate, engulfed in the flames of apostasy, intermarriage, and assimilation. This condition was, moreover, exacerbated by the presence of both political and racial antisemitism. So it was in Germany that the task before the Zionist anthropologists was most complex. They girded themselves for battle with an enemy that attacked from within and without.

Whereas Joseph Jacobs had observed the Russian pogroms from afar, and Samuel Weissenberg contemptuously viewed German Jewish assimilation from Elizavetgrad, German Zionist anthropologists bore personal witness to the effects of both on their own soil, envisioning the inevitable disintegration of their community.[3] They could not, as Jacobs had done in England with the Russian Jews, encourage them to acculturate and partake of the fruits afforded by a liberal, democratic nation. And for a large percentage of urban Jews in Central Europe, acculturation had blossomed into assimilation. Indeed, to use the felicitous phrase of one historian, a significant proportion of German Jews had by the first world war gone "beyond Judaism."[4] As far as discrimination was concerned, occupational restrictions as well as barriers against full social integration, though not unbearable for the majority, blocked the way of many an aspiring Jew.[5] And unlike the antisemitism that Weissenberg knew was the lot of Russian Jews, by the last quarter of the nineteenth century a more insidious, uncompromising variety based on irrevocable racial difference confronted German Jews.[6]

The most radical Zionist critique of these circumstances, and indeed the last two thousand years of Jewish history, entailed nothing short of a total rejection of continued Diaspora existence.[7] Thus, for the radical Zionists, there was but one thing for Jews to do: actively bring about the end of exile.[8] There could be no future for Jews on a continent where they would continue be a reviled minority. Furthermore, the condition of exile meant the Jew was without roots, and in the view of the Zionists this only served to promote the process of spiritual atrophy and physical decay. In stark contrast to Samuel Weissenberg, hard-line Zionists denied that the Diaspora helped give the Jews their strength, enhanced their vitality, and fostered Jewish civilization's tenacious longevity.

To bring about the success of their cause, the Zionists regarded the creation of a new collective identity for the Jews as essential. For many Zionists who redefined Jewishness in nationalistic terms, the concept of race was an alluring device. This is especially the case for a movement that was at pains to prove not only to the outside world but also to Jews that despite the heterogeneity of language, custom, and physical appearance, they were, in the words of the Zionist leader Theodor Herzl, "one people."[9]

But the use of race science by the Zionists was not a mere tactical ploy. Their acceptance of this model of ethnic differentiation was in part a consequence of their own social backgrounds. Paradoxically, even though the Zionists saw no future for the Jews in Europe and sought their mass exodus to Palestine, they were nevertheless, in the case of Central Europe, committed Europeans. As Robert Weltsch, the postwar editor of the Zionist *Jüdische Rundschau,* wrote of his generation, "what was important . . . was not the farewell to Europe, but the greedy acceptance of all that Europe had to give us." To put a finer point on it, even for the German Zionists, "Europe inevitably meant Germanness."[10] In addition, most of the movement's leaders were financially secure and belonged to either the middle or upper classes. For Wilhelminian Jewry, this was often synonymous with almost total estrangement from Jewish culture and a complete affiliation with German *Kultur.*[11]

One facet of this adopted cultural legacy was racialism. At the turn of the century, Europeans of all political and cultural backgrounds believed in the concept of race. The emphases they placed on and the lessons they drew from the dogma may have varied, but the proposition that essential differences existed between various groups of people was not seriously questioned by the majority. The Jews, including the Zionists, were part of that majority.

With the Zionist anthropologists, the marriage of science and ideology was as steadfast as it was with their bitterest enemies, the racial antisemites. Whereas the latter most often stressed the inferiority of the Jews (and at other times the great threat caused by their superiority), the Zionists battled to assert that the Jew was not a member of a degenerate race but, on the contrary, a member of a people on the threshold of national rebirth.[12] It is a paradox—and German Zionism is full of them—that before bidding farewell to Europe, Zionist race scientists, like their liberal Jewish contemporaries, sought to demonstrate the invaluable contribution Jews had made to the development of European civilization.[13]

But despite limited points of agreement such as this, the Zionists' battles with their coreligionists could be fierce. The Zionistische Vereinigung für Deutschland (ZVfD), the main Zionist organization in Germany, mounted a counterattack not only against the antisemites—the majority felt the battle was unwinnable because antisemitism was a necessary condition of Diaspora existence—but also against the liberal Jewish establishment, which asserted that Germanness and Jewishness were not incompatible. According to this well-known formulation, the Jews were to be regarded merely as Germans of the Jewish faith, just as there were Germans of the Catholic and Protestant faiths; all, however, were unquestionably German in a cultural sense. This was the view expressed by the Centralverein deutscher Staatsbürger jüdischen Glaubens (CV), the umbrella organization that represented the majority of German Jews by the mid-1920s.[14] The Zionists were often wont to hold the CV responsible for the prevalence of intermarriage and conversion, but in truth the CV was just as opposed to these community-debilitating phenomena as were the Zionists.[15]

Thus in Germany, Jewish race science, as practiced by Zionist anthropologists, was more than a knee-jerk reaction to racial antisemitism. Above all, it was employed within Jewish circles as a weapon in the political battle for the hearts and minds of German Jews. With that end in mind, one can see how Jewish racial science developed in the true tradition of Wissenschaft des Judentums, the early nineteenth-century movement for the academic study of Judaism. Its stated goals were not merely the objective study of the Jewish past through critical analytical methods and the achievement of a respected place in German universities, but also to inculcate a sense of pride among those Jews who had become estranged from or, even worse, embarrassed by their own people and heritage.[16] So in Germany, Jewish racial science was intended for both internal and external consumption. It is against the background of the special inner tensions and external exigencies brought about by large-scale intermarriage, radical assimilation, Jewish nationalism, and German antisemitism that we must analyze the work of the foremost German Zionist physical anthropologists.

Elias Auerbach

In 1907, an obscure twenty-five-year-old doctor from Berlin, Elias Auerbach, made his anthropological debut with an essay in the prestigious *Archiv für Rassen- und Gesellschafts-Biologie*, in which he addressed the question of Jewish racial purity.[17] Auerbach's work attracted the attention of established anthropologists to the extent that it did because in Germany, no one before Auerbach had made such a sustained and detailed case for the racial purity of the Jews.

Unlike many of the leaders of the German Zionist movement, Auerbach was raised in a pious environment, in the province of Posen, where he was born in 1882. After receiving a traditional Jewish education, he attended the Königstädtische Gymnasium in Berlin and went on to take his medical degree in 1905 at that city's university, which at the time had a large medical faculty of some five hundred students who were instructed by some of Europe's most celebrated professors of medicine.[18] While in medical school he read every day from Spinoza and the Bible; throughout the life and work of this eminent Bible scholar and historian,[19] science and skepticism would sit easily with tradition and faith.[20]

It was also during this period that Auerbach became a Zionist. His sister Johanna had married Heinrich Loewe, a Jewish folklorist and librarian and one of the founders of German Zionism. In 1901 Auerbach moved in with them, and through Loewe's instruction he became actively engaged in the Zionist cause. In 1905, thoroughly committed to the establishment of a Jewish national homeland, Auerbach became one of the earliest of the German Zionists to move to Palestine, where he resided in the port city of Haifa until his death in 1971.[21]

For Auerbach, the racial question was "the question of the constituent parts of a race" (332). In other words, what were the historical factors that gave rise to miscegenation, and how had the process played itself out in the physical appearance of races? Specifically, in order to understand the Jews in racial terms, Auerbach insisted that three essential points had to be clarified: "(1) In the 'historical' period, had no appreciable intermixtures taken place?; (2) In prehistoric times, had demonstrable intermixtures occurred?; and (3) How are we to conceive of the original Jewish race [*Urrasse der Juden*]?" (333).

Following the methodological precedent set by Joseph Jacobs and continued by nearly all subsequent Jewish race scientists, Auerbach built his anthropological case upon the foundation of historical evidence, albeit highly selective and overly romanticized. For him, history became anthropology's handmaiden: "These two ways of looking at things, the anthropological investigation of the living race and the historical identification of its wanderings and transformations, must go hand in hand." As far as this pertained to the Jews, Auerbach noted with some pride that "that is why the Jews are a classic object of racial research, because we can permit history and measurement to work together [with them] better than we can with any other race." He asserted that when history and anthropology were brought to bear on the study of the Jews, they revealed a race in which, "in the period commonly known as the 'historical,' hardly any more intermixture took place" (332–333).

Auerbach's study consisted of a wide-ranging historical survey of the biological encounters between Jews and gentiles throughout the course of Jewish history. Beginning with the Canaanite period, and continuing through to the theme of intermarriage in his contemporary Germany, Auerbach sought to prove that throughout the bulk of their history, the Jews had remained a pure race, unaffected in any appreciable way by intermixture with other peoples, and that "they demonstrated more clearly than any other race how overpowering is the influence of heredity, in contrast to that of adaptation, for the destiny of the race [*Rassenschicksal*]" (333).

In the presentation of his historical evidence, Auerbach worked his way backwards from the present. As a Zionist polemicist, he was above all concerned with the crisis that confronted the contemporary Jewish community in Germany, in which by 1903 intermarriages accounted for one-sixth of all marriages involving Jews. Approvingly quoting the evidence accumulated by the Zionist statistician Arthur Ruppin, Auerbach lamented that "the number [of intermarriages] is so great that from this alone one must conclude that German Jewry will soon be completely liquidated" (334). But the social consequences of Jewish intermarriage were not his prime concern and would be taken up in greater detail by another Zionist physician, Felix Theilhaber.

Auerbach's main goal at this point was to determine the extent to which the offspring of these marriages deviated physically and biologically from the original Jewish racial type. He claimed it as indisputable that the vast majority of these children of mixed marriages grew up totally outside the pale of Jewish religious and national life and therefore were inconsequential as far as the question of "race mixture" was concerned. Intermarriage was thus a social but not a racial problem. Of course, at a time when falling birthrates were of paramount importance to nationalist social critics (some of them Jews), large-scale intermarriage obviously translated into a real demographic loss for Jews. But in an increasingly eugenically-minded world, quality came to count at least as much as quantity. For Auerbach, whose primary concern was to establish Jewish racial purity—in his view, a mark of quality—intermarriage did not, as he had shown, constitute a threat to this status. On the racial integrity that the Jewish people enjoyed, Auerbach remarked that they had "remained a closed mass," with only "few changes" having taken place. In fact, he boldly asserted as his fundamental premise that "in the entire Middle Ages—and for Jewish racial history the concept of the Middle Ages extends to the French Revolution—the number of mixed marriages was so infinitely small that we can completely neglect them, especially since there was hardly one case in which the sons of such mixed marriages further mixed their blood with the Jews" (334).

According to Auerbach, the biological isolation of the Jews over so long a period was partly a result of the economic role they played in the various societies in which they lived. Because the same picture of Jewish separatism appeared all over the world, Auerbach presumed that a universal historical process was in operation whereby Jews traditionally either entered sparsely populated and poorly cultivated lands of their own accord or were invited in as guests. Bringing with them their more sophisticated systems of commerce, the Jews rapidly formed a middle-class wherever they went. Soon the economic station of the surrounding peoples was sufficiently elevated, and they then felt they could dispense with the Jews, whose presence only led to harsh competition. The process of exclusion was conducted by the imposition of special laws—in seventh-century France, eleventh-century Germany, thirteenth-century Poland, and fifteenth-century Spain—which Auerbach maintained were designed to isolate the Jews. Certainly after the Spanish expulsion (but probably long before it), no racial intermixture between Jews and non-Jews took place until the nineteenth century (335).

This economic analysis of Auerbach's, simplistic, naive, and overly deterministic as it was, had long been a stock-in-trade argument of the European left and even recalls certain centrist and right-wing objections to Jewish economic activity, with their notions of Jewish responsibility for the development of modern capitalism. Four years after Auerbach presented his thesis, the German political economist and sociologist Werner Sombart published his notorious and controversial *The Jews and Modern Capitalism* (1911). Focusing on the effects of intense competition between medieval guilds and the Jews, Sombart likewise charged the Jews with responsibility for the development of more advanced systems of international commerce, which according to him resulted from the Jewish desire to break free of the restrictive commercial environment created by the city-administered guilds. In language reminiscent of the scenario of Jewish settlement laid out by Auerbach, Sombart wrote, "When Israel appears upon the face of Europe, the place where it appears comes to life; and when it departs, everything which had previously flourished withers away."[22] In Sombart's conspiracy theory, the Jews single-handedly control Europe, determining the fate of its totally passive Christian inhabitants.

Quite clearly, the aims and methodologies of Auerbach and Sombart were totally at odds. The former was trying to prove a biological point by recourse to economics, and the latter was attempting to create an economic theory by recourse to a biological or racial argument.[23] Yet the two drew from the same well of explanations to make their points. How often the arguments of Zionists and antisemites, as the early Jewish opponents of Zionism declared, sounded uncomfortably similar.

Modern readers will despair of fin-de-siècle race science texts, irrespective of authorship, if we seek consistency of argument. Very often a scientist will contradict himself from one work to the next, or even within the one study. To practice race science was to embark upon an intellectual journey whose route entailed negotiating a slippery slope. In fact, consistency was to be found mostly in the writings of the most blatantly racist of authors. They could hammer home a point which, though illogical, would be unambiguously and tenaciously argued. The race scientist who sought objectivity, the one who attempted to grapple sincerely with questions of nature and nurture and of morphological difference, was the one most likely to waver, display doubt, and be self-contradictory.

This happened to Auerbach. His discussion of the Jews in Muslim Spain in his study of the Jewish racial question epitomizes the precariousness of race science. Suddenly his argument is less environmentalistic than his previous analysis of economic development had been, and takes on a far more biologically deterministic coloration. Delving into the murky depths of *völkisch* notions of inherent racial instinct and feelings, Auerbach credited the Jews of Spain with having had the most highly developed sense of ethnic uniqueness and biological destiny of all premodern Jewish communities. He was unable to deny that intermarriage had taken place in Spain, where the favorable social and economic conditions that both Jews and Muslims enjoyed made the probability of intermixture rather high. But Auerbach maintained that there was an essential difference between intermarriage as it had occurred in Spain and that which was presently taking place among the Jews of Germany's largest cities. In rather vague terms, he concluded that the main difference lay in the fact that in the earlier epoch "Jews showed no inclination to abandon their racial isolation." He went on to make a more general statement, that "in the course of their entire racial history it has been the Jews themselves and not the other peoples who have promoted the strongest resistance to racial mixing" (335).

Clearly contradicting his earlier postulation that economic competition had led the non-Jewish authorities to enforce the separation of Jews from the rest of society, Auerbach declared that the true source of this Jewish resistance was to be found in the religion itself. The acknowledged racial difference of Jews was codified and enshrined in a body of laws whose social control mechanisms served to isolate them. According to this view, *halakha*, or Jewish law, became the decisive element in the preservation of the race. Auerbach was not alone in holding this view. Nearly all race scientists, Jewish and gentile alike, lauded what they regarded as the racially positive effects of Jewish law. While some focused specifically on the supposed salutary effects of the dietary laws, others stressed the morally rewarding benefits of the Jewish code of sexual conduct. Still others, often even rabid opponents of the Jews, approvingly cited Jewish prohibitions against intermarriage as proof that the Jews had, since their advent, been the most eugenically-minded of all peoples. Their continued survival was the clearest proof of the efficacy of all these regulations. While other peoples had disappeared during the course of history—presumably because they lacked the racial drive to remain separate—the Jews, thanks to their instinct of self-preservation, had survived. In southern Spain the region's two dominant religions, Judaism and Islam, were so closely related that the Jews refused to accept proselytes so as to prevent the physical mixture with another race.

So for Auerbach, the Jews of Spain, unlike those of modern Germany, were guided and protected by an acutely developed racial instinct, bolstered by strict adherence to religious laws that he read as precluding the biological admission of gentiles into Jewish ranks. The lesson he drew from a comparison of medieval Spanish Jewish and modern German Jewish marital behavior was that medieval Spanish Jewry represented an acceptable model of conduct for contemporary German Jews. According to Auerbach, Spanish Jews provided an example of an acculturated Jewish community that still preserved intact its separatist identity, both cultural and racial, whereas German Jews had become assimilationists with no instinct for self-preservation, destined to blend insensibly into the German nation and disappear.

Auerbach, and Jacobs before him, largely constructed arguments about Jewish racial purity on the supposition that Jews had historically rejected the recruitment of non-Jews to their ranks. The issue of gentile converts to Judaism assumed a central place in turn-of-the-century debates on the Jews as a race. For proselytism, if it had taken place on a large scale, would explain the so-called physical anomalies among the Jews—blue-eyed blondes, redheads, and Jews with African and Asian features, as well as the perplexing predominance of brachycephalism among world Jewry. (Recall that Samuel Weissenberg built an entire theory of Jewish brachycephalism on the basis of his belief in the Khazar conversion of the eighth century.) Trying his hand at an explanation, Maurice Fishberg, the celebrated American Jewish race scientist, offered that the existence of the different anthropological types among the Jews was the result of the twin phenomena of proselytism and the birth of children to Jewish women raped during centuries of pogroms.[24]

By the late nineteenth century, the evidence that numerous scholars had accumulated concerning the history of proselytism and the legal and social status of the Jewish proselyte overwhelmingly contradicted the myth (still popularly maintained to this day) that Judaism did not seek out converts. Although proselytism was not actively pursued in ancient Israel, both Josephus and the Latin writers attest that in the post-exilic period a large number of pagans converted to Judaism. Especially after the destruction of the second temple and through the sixth century CE, many rabbis openly encouraged it, subscribing to the dictum *karev takhat kanfei ha-shekhinah*, "bring them under the wings of the Divine Presence."[25] From the time of the Council of Orleans (538), legislation was passed prohibiting Jews from converting their pagan slaves to Judaism, but voluntary conversions to Christianity continued to take place. Naturally, proselytism declined throughout the Middle Ages with the worsening situation of the Jews and the concomitant rise of the mendicant orders.[26]

Auerbach simply ignored much of the historical evidence available to him. In attempting to minimize the biological impact on the Jews of a significant number of converts, he categorized most of those who had converted under the Romans as partial proselytes, known as Proselytes of the Gate. This status meant that the group accepted the principle of monotheism and Jewish ethics, observed the Sabbath, and refrained from eating pork, but did not submit to circumcision or practice other ritual ceremonies. Asserting that these were the majority of converts and that they had left no trace of themselves on the Jews, Auerbach declared that "with Titus's destruction of the Jewish state, there began the ... self-preservation struggle of the Jewish race, [and] on the whole one can say that in the last two thousand years the Jews have practiced the principle of racial inbreeding [*Rasseninzucht*] with great tenacity. Throughout they have secured the inheritance they brought with them out of Palestine."[27] In other words, contrary to the historical evidence of widespread conversion and to religious custom that allows for the full participation of the proselyte in Jewish life, Auerbach maintained that Judaism knowingly discouraged potential converts, thereby preserving Judaism's religious and biological patrimony.

Many hostile critics in the medical profession, particularly in psychiatry, charged that the Jews displayed a proclivity to physical and psychological degenerative diseases because of intramarriage or, as they most often put it, *Rasseninzucht,* "racial inbreeding." They based their theories on the same assumptions as Auerbach about Jewish racial purity and aversion to intermarriage and proselytism. They regarded this tendency of Jews to marry other Jews negatively and considered it the cause of numerous illnesses believed to be peculiar to Jews. But Auerbach and other Jewish raciologists, such as Arthur Ruppin, saw Jewish intramarriage in an entirely positive and beneficial light.[28] Its practice was a tangible expression of a sound racial instinct and will to survive. Rather than causing degeneration, it had brought about the healthy regeneration of the Jewish people for two millennia. It is ample testimony to the malleability and value-laden enterprise of race science that terms such as "racial inbreeding" were not automatically accusations of impropriety. They remained loaded, but were only damning once a particular author's anti-Jewish polemical intent could be ascertained.

In writing a racial history of the Jews from a Zionist perspective that denied the fact of large-scale conversion to Judaism, Auerbach sought to trace an unbroken chain of Jewish biological transmission from biblical times to the present. His claims of Jewish racial purity and his advocacy of the theory that modern Jews were the direct descendants of the twelve tribes meant that the return of Diaspora Jews to their original homeland would constitute a historical phenomenon that was in perfect harmony with the biological nature of the Jewish past as Auerbach had read it. Zionism, in its rejection of the classic Christian bifurcation of Jewish history into ancient Israelite (noble) and postbiblical Jewish (ignoble) periods, stressed the continuity of Jewish history, thereby emphasizing the national idea and national destiny. Furthermore, Zionist historians saw the Jewish nation as fundamentally different from other nations because of this perceived historical continuity. The point was made clear by Joseph Klausner, editor of the Hebrew monthly *Hashiloah*: "A deep chasm separates the ancient Romans from the Italians, the ancient Hellenes from today's Greeks: religious beliefs and even the names of the nations have changed. But the same has not happened where the Jews are concerned ... they never lost their awareness of the integrity of their national personality."²⁹ Auerbach's Zionist view of Jewish history contained within it two competing methodological and ideological paradigms, both of which are inherent to revolutionary nationalist movements: continuity and change. For the Zionists, this problem meant creating a new national Jewish identity based on the traditional concepts of common origin and shared historical experience while at the same time attempting to effect a break with the post-exilic past.

Thus, despite the conservative rendering of Jewish history (albeit through the extremely narrow prism of the people's biological status), Auerbach's interpretation also carried with it the promise of historical rupture. With the glory of the Jewish past threatened by the people's potential demise, Zionism's clarion call would awaken the Jewish masses from the soporific effects of assimilation. Zionism, rather than an organic part of normative Jewish historical development, was for Auerbach an invention designed to save Jewish history through its powers of revival and rebirth. It would, according to an optimistic Auerbach, serve as a catalyst to bring out the best of what had recently become dormant in the people. As he wrote in a Zionist student publication, "For us, the justification for the existence of the Jewish people results solely from our conviction that today, cultural power still lives within it. But this power lays bound, bound in homelessness and social misery. The liberation of this power from out of these binds is served by that movement which we call Zionism."³⁰

Unlike many Zionists, Auerbach did not see exile as an insufferable period of prolonged agony. Instead, he portrayed it as a time of test for Jews—a test they had passed with flying colors. For two thousand years they had kept the outsider at bay for the conscious purpose of retaining their original and ancient (*uralt*) racial characteristics. This points to yet another deviation from "mainstream" Zionist ideology, one which I believe was unavoidable for Auerbach and the other Zionists who wrote scientific apologia. As one historian of Zionism has observed, a central motive behind the movement was the "retrieval of Jewish dignity." Its loss had been gradual, but by the end of the nineteenth century was almost complete. Jews in the West, ashamed of their Jewishness, had come to a state where "the consistent humiliation and slander of the Jews has led us to begin believing these lies."³¹ But, by contrast, Auerbach's interpretation of the Jewish past reads as a hommage to the people's dignity. Nowhere did they betray their racial origins, at no time throughout their wanderings did miscegenation threaten the existence of the race. For Auerbach, the Jews' drive for "self-preservation" had been fueled by an innate

135

sense of pride and dignity, and thus the history of the Jewish diaspora was cause for pride and celebration.

But two millennia was an extremely long time to mount such a spirited defense. The Jews, particularly those of his Germany, had since the beginning of the nineteenth century begun to falter. The path of assimilation was leading them to racial oblivion; soon they too would disappear like so many of the peoples they had outlived during the course of their long history. But a solution to the dilemma was at hand. In a theory pregnant with teleology, Auerbach postulated that just as the Jews were beginning to fail, Zionism, the most dynamic of Jewish ideologies, would be their savior. Zionism's mission of returning the Jews to the land of Israel would, in effect, place them back into the position they enjoyed before the nineteenth century—politically autonomous, culturally whole, and racially pure.

Auerbach's hypothesis, built on the mystical premise that there existed a Jewish racial instinct whose effectiveness had ensured racial exclusiveness and therefore Jewish racial purity, was reminiscent of much of the German *völkisch* literature being disseminated by nationalist (and often antisemitic) groups. Zionism, as expressed by the influential German-Jewish students' organizations, of which Auerbach was a leading member, often used classic *völkisch* terms such as "will" and "battle" to describe the historical path and destiny taken by the Jewish people. As one student summed it up, "In a sick Jewry Zionism is the will to live . . . we rejoice in this battle which signifies life."[32] Or as Auerbach himself declared with reference to the clash between German and Jewish nationalism as expressed at the 1902 conference of the Deutschen Freien Studentenschaft, "Will against will. And the will to live will be more powerful than the will to go under."[33] By speaking in such terms, the Zionism of the educated elite in Germany was far removed from the atavistic individualism of the liberal ideology favored by the majority of German Jews. The language of the Zionists permitted individual Jews to feel as if they had become subsumed into the nation, participants in a great national experiment. Asserting the collective "will" and doing "battle" provided youthful Zionists with a positive self-image, one in which they could consider themselves members of an ancient, national-historical, or racial fraternity.

Where did this primitive racial instinct, this "will to live," develop? For the Zionist Auerbach, the answer was obvious: Palestine. Any attempt to write a racial history of the Jews and determine their relative racial purity entailed taking the debate over racial intermixture to the source. In doing so, Auerbach became embroiled in the controversy over Jewish racial origins and the issue that so dominated the work of Samuel Weissenberg, namely, the character of the Jewish *Urtypus*. This in turn meant directly confronting and refuting the theories of Felix von Luschan.

The dispute centered on von Luschan's claim that already in Palestine the Jews had mixed with the non-Semitic, brachycephalic Hittites and the blond Amorites. For von Luschan, it was this ancient crossing that accounted for the prevalence of broad-headedness and fair complexions among modern Jews. For Auerbach, who was an Orthodox Jew, the Bible furnished nearly all the evidence required to prove that in the land of their national birth the Jews had remained essentially pure. Commenting on the biblical period, Auerbach proudly declared: "With a racial pride that, according to all evidence, is possessed in great measure by the Jews, it should not astonish us to meet with regulations against the penetration of a foreign type into the race. A comparable example is offered by the conduct of the nobility, whose exclusivist outlook corresponds to the time of the least mixing of 'inferior' blood" (338).

Relying almost entirely on the biblical accounts of Israelite marriage, with only occasional recourse to archaeological evidence concerning the geographical dispersion and domicile of the various races that made up the indigenous population of the land of Israel, Auerbach denied that the ancient Israelites could have mixed with foreign peoples and affirmed his belief in Jewish racial purity. As far as the considerable racial crossing that von Luschan maintained had taken place between the Jews and the Hittites, Auerbach rejected it on the assumption that when the Israelites appeared, the Hittites no longer occupied large portions of the land.

In sum, Auerbach found four compelling reasons for rejecting the Hittite hypothesis: the silence of the biblical sources on the Hittites; the insignificance of Hittite language and culture on the Jews; the exclusive derivation of contemporary Jews from the tribes that dwelt in the south of the country, whereas the tiny Hittite colonies were in the north; and the probability that the Hittites were hyperbrachycephalic, whereas the Jews are chamaebrachycephalic, or slightly less that brachycephalic. Auerbach simply concluded that "the explanation that the short-headedness of the Jews is caused by intermixture with the Hittites in Palestine runs into the greatest difficulties" (346).

As far as Jewish intermixture with the Aryan Amorites was concerned, Auerbach first rejected the commonly held premise that the Amorites were dolichocephalic blonds. In 1888, some time before Auerbach had made this declaration, two other anthropologists had cast severe doubt on the theory. Measurements on the Egyptian mummy of Thutmes III by the renowned pathologist Rudolf Virchow had proven that even a skull that appeared on sight to be quintessentially dolichocephalic, as had that of the pharaoh, could turn out in fact to be short-headed. As for complexion, the British anthropologist Flinders Petrie had claimed that according to pictorial evidence, the Amorites had "red-brown hair."[34] As he had done with the Hittites, Auerbach now said of the Amorites, "The explanation that the blondness of the Jews is caused by intermixture with the Amorites runs into the greatest difficulties" (348–349).

But the significant population of blond Jews also had to be accounted for. During the last quarter of the nineteenth century, race scientists examined the pigmentation of approximately 153,000 Jews all over Europe. Some 145,000 of these subjects were schoolchildren, and the remaining adults were categorized according to location, sex, and whether they were Sephardim or Ashkenazim. While the figures confirmed that Jews were predominantly dark-haired, the percentage of blonds in some places was noticeably high.[35] As to be expected, the figures did not ensure anthropological consensus, and a number of conflicting explanations were offered. Prichard and Jacobs, along with the Frenchman Pruner Bey, attributed the phenomenon of blond Jews solely to the influences of climate and environment. Other scientists, such as Broca, Lagneau, Virchow, and Schimmer, claimed that the cause was intermixture with Aryan elements in modern times.[36]

For Auerbach, a more complex mechanism of the transmission of this particular physical feature was in operation. According to him, the Jews had long practiced a process of sexual selection, whereby the physical features of the surrounding population had become their ideal type. Where, for example, a region's majority population had a high percentage of blonds, as in Germany, so too did the Jewish population there. These conditions led to a "physical, and not only cultural, assimilation of the Jews": when it came time for marriage, Jews showed an unconscious preference for blonds, the *Schönheitsideal* of the gentiles among whom they lived. Auerbach regarded this selection process as progressive in that the blond elements were multiplied, and since such persons had a greater chance to produce offspring the demographic benefits to the Jews would accrue (358–359).

Von Luschan's theory also took into account the physical intermixture between Jews and the Semitic peoples of Canaan. On this point the two anthropologists were in accord. Conceding that some mixture with non-Jewish races had occurred in Palestine, Auerbach asserted that it had mostly been with members of other Semitic races who were linguistically, culturally, and physically similar. And in anticipation of an obvious criticism, the author quickly noted that there was no evidence to question the original purity of these other Semitic races (339).

Once he had decisively refuted all claims of substantial Jewish commingling with foreign peoples, it remained for Auerbach to explain the physical anomalies without recourse to a theory of racial crossing. This particular challenge led him to overturn universally held anthropological theories concerning the Semitic peoples. Striking at one of the most sacred beliefs of modern craniometry, that the Semites were originally dolichocephalic, Auerbach claimed that the *Ursemiten* were brachycephalic or at least inclined to short-headedness. This was a sine qua non of the syllogistic hypothesis that ran: Nearly all Jews are brachycephalic; all Jews are Semites; Semites must therefore be brachycephalic. "It is really to be doubted whether the original Semites are to be characterized by their small, long heads. If we recognized their long-headedness, it would follow that, according to the present measurements of the Jews, they have at most 10–15 percent Semitic blood, while the 85–90 percent whose distinguishing feature is short-headedness, or broad mesocephaly, is to be attributed to intermixture. But that would mean that generally the Jews are not to be considered Semitic, and that the Semitic in them is to be considered an incidental racial infusion. One would be justified in protesting against this interpretation." Just what this meant for the Jews, and for their relations to the peoples among whom their civilization developed, Auerbach made abundantly clear. He declared "the true unity of character of the Babylonians, Assyrians, Phoenicians, Arabs, and Jews. It is a unity that is made known in their illustrations, their language, their legends, their architecture, and frequently in their original religious forms, down to the smallest details. The pictorial representations of Assyrians and Babylonians that we have show not only a great conformity with the ancient representations of Jews but also a surprising similarity with today's Jewish type. We are also justified in speaking of a general Semitic type, to whom the Jews belong (351)." As a Zionist who sought the departure of the Jews from Europe, Auerbach's expression of fraternal bonds between Arabs and Jews carries with it clear, and what were for him hopeful, political implications.

By recourse to such arguments about hair coloration and skull shape, Auerbach was able to salvage his hypothesis of Jewish racial purity. But his zeal in defense of that theory shows that his Zionism impinged on his science. By claiming that a unique and ancient bond existed between all the peoples who spoke Semitic languages, Auerbach was asserting the appropriateness of a mass Jewish return to the Middle East. Buoyed by ample anthropological evidence and by theories of Semitic unity, Auerbach's Zionist vision projected a peaceful and harmonious future for Jews and Arabs in the Land of Israel. The hallmarks of this new society would be mutual respect and cooperation, albeit under Jewish control. He made this point abundantly clear in his warmly received book *Palästina als Judenland* (1912). There he urged that Jewish settlers not expropriate Arab land but rather work for the mutual benefit of the two peoples. He warned that while it was "necessary and indispensable for the Jewish element [to have] a firm and conscious national unity," it must not develop "a provocative chauvinism."[37] For example, when addressing the issue of what the official language of the Jews in Eretz Yisrael would be, an issue of seminal importance for the early Zionists, Auerbach was in a minority when he stated his hope that "the time is not so far off" when the two official and dominant languages would be Hebrew and Arabic.[38]

The harmony he envisaged for Palestine was hardly based on a history of smooth relations between the Jews and the peoples of Europe. He had shown the former to have been a Semitic people who had remained physically unchanged, and thus essentially different, from the Europeans among whom they lived. Thus, although the Jews had to leave Europe, where ancient racial antagonisms had either made life unbearable or, because recent assimilation, would result in the demise of the Jewish race, Jewish redemption in the Land of Israel, where Jews would live among those to whom they were racially and culturally akin, was in the natural order of things.

Von Luschan wasted little time in responding. In an open letter he challenged Auerbach's argument point by point, accusing him of being unscientific, of not having made the requisite field trips—especially to the Near East, as he himself had done—and for generally being a better biblical scholar than race scientist.[39] Von Luschan was of the opinion that there was only a Jewish religious community and no such thing as a Jewish race. Subscribing to the majority view among modern anthropologists, he asserted that race mixture (as opposed to purity) was a key to the higher ranking and future development of certain races, such as Jews.[40] In this same letter, von Luschan challenged the Zionist premise of Auerbach's anthropology and voiced his stern opposition to the Zionist movement:

> In my lecture of 1892, I concluded with the sentence that our Jewish citizens are our best and truest co-workers in the struggle for the highest good on

this earth, in the struggle for progress and for intellectual freedom. I stand today essentially by the same point—with one certain qualification: those who are connected with the Zionist endeavor. I regard these people, who appear to have won some ground, as direct enemies of civilization [*kulturfeindlich*].... For us today, the main attraction of a journey to the Orient is the absolute medieval atmosphere that surrounds us there and the possibility of attaining, within a few days in this environment, a sense of what ours was like a millennium ago. In addition to the real enjoyment of immersion in this thousand-year-old past is the consciousness of always being able to return to modern civilization. A lengthy sojourn in the Orient would appear to many of us only little better than a stay in a penitentiary. And so I think that a repatriation of even only one part of European Jewry to the Orient, for the time being, would signify a retreat into medieval barbarism. But already those from various sides are now animated to create a union of all Jews, and the attempt to anxiously protect them as a "pure race" from all intermixtures with their neighbors is hostile to civilization. In part, the attempt is based on that certain arrogance that is characteristic of the Orthodox Jews, and in part on the view of the "racial purity" of the "chosen people." To refute this view and to constantly prove the new admixtures out of which the Jewish people of today has arisen is, therefore, of practical significance.[41]

Von Luschan had clearly identified how Auerbach's anthropology was underpinned by his Zionism. For all its scientific pretensions, Auerbach's race science was deeply flawed, having been enlisted to serve as the ideological premise of a national movement. But irrespective of who the practitioner was, race science at the fin-de-siècle was always allied to one ideological position or another—von Luschan's no less than Auerbach's. The dispute between the two men went on long after race science had become a thoroughly discredited enterprise. In a stunning admission made as late as 1965, Elias Auerbach had the final word on the subject. Reflecting on his early anthropological work, he wrote, "When I read it ["Die jüdische Rassenfrage"] again today, after 60 years, I do not have the impression that it is outdated."[42]

Felix Theilhaber

In 1911 a book appeared that sent shock waves through the German Jewish community, for it foretold the decline and eventual disappearance of that group. Apocalyptic stories about the demise of races, especially the Jews of Germany, were not new. After all, Houston Stewart Chamberlain's *Foundations of the Nineteenth Century*, with its visions of combat unto death between Aryans and Semites, was already a dozen years old. What was unique about this new work, with its alarmist title, *Der Untergang der deutschen Juden* (The demise of German Jewry), was that it spoke not of race wars but race suicide, and was written not by an antisemite but by a Zionist physician named Felix Theilhaber.[43]

Born in Bamberg in 1884 to a well-to-do family with only the loosest ties to its Jewish identity, Theilhaber was raised in Munich, where his father, Professor Adolf Theilhaber, enjoyed a celebrated career as one of Germany's foremost research gynecologists.[44] The young Theilhaber took up the study of medicine at his father's insistence, becoming not only a successful dermatologist but also an indefatigable writer and political activist. The list of his published works ranges across many genres and includes medical monographs, war memoirs, Jewish apologetics, novels, a study of Goethe, and Jewish history.

Theilhaber took an active interest in politics. He became a committed Zionist around 1900, and in 1901 helped found a Jewish gymnastic society in Munich. As a university student, he also founded the Jewish fraternity Jordania, the Kartell zionistischer Verbindungen, and the Herzl-Bund for Zionist merchants. In 1906 he went on his first sightseeing trip to Palestine. He later recalled this watershed event in his life rather romantically: "I had seen the land of my Fathers. It was full of stones and not one cedar, but for me, Eretz Yisrael was no longer an empty concept, but a reality, a piece of earth whose magic power I was never able to shake off." Upon his return, Theilhaber traveled to Eastern Europe to enlist support for the Zionist project from the powerful Belzer rebbe. He was granted an audience, but the rabbi denied his blessing for the movement and its aims, telling Theilhaber in no uncertain terms that the Jews would remain living together with the Poles until the Messiah arrived. One year later, in 1907, Theilhaber established a newspaper called *Palästina*, of which he served as publisher and editor until 1910.[45]

Theilhaber's political activities extended beyond his dedication to the Zionist cause. In 1913 he founded the Gesellschaft für Sexualreform, whose twin aims were the legalization of abortion, meaning the repeal of paragraph 218 of the criminal code, and the easing of public access to contraceptives.[46] Theilhaber also worked closely with the powerful Bund für Mutterschutz and its founder, Dr. Helene Stöcker. Later on, as the movement grew in strength and influence, Theilhaber's organization joined the Reichsverband für Geburtenregelung, and under the leadership of Theilhaber and the Reichsverband's Dr. Schöffer it published the important *Zeitschrift für Sexualhygiene.* They were joined in the enterprise by the Berlin Jewish physician and sexologist Magnus Hirschfeld.[47] In 1935, as victims of Nazi persecution, Theilhaber and his family fled Germany and settled in Palestine.

Theilhaber's activities on behalf of sexual reform and his Jewish background and identity are perhaps not so far removed from each other as would first appear. He began his medical practice in Berlin as a dermatologist and specialist in sexually transmitted diseases. At the time, this branch of the medical sciences carried little prestige and consequently attracted many Jews who had been shut out of the more traditional specialties because of their Jewishness. Moreover, the campaign of Jewish doctors such as Theilhaber and Hirschfeld for the rights of homosexuals and women as well as those suffering from venereal diseases—all groups who at the turn of the century were labeled as outsiders—was related to their Jewishness. These men also felt themselves to be outsiders—professionally, because as Jews they were pushed to the periphery of medical science, and individually, Theilhaber because of his Jewishness, Hirschfeld because of that and his declared homosexuality.

Theilhaber graduated from medical school in 1910. His doctoral dissertation drew on his father's research interest in uterine cancer but also presaged the methodological approach he would soon adopt in his own statistical research on Jews. Combining medicine and sociology, it established that the incidence of cancer of the uterus among Jewish women was considerably lower than among Protestant and Catholic women. The reasons for this, he maintained, were social and not racial, mostly having to do with Jewish sexual ethics and bourgeois status.[48] The following year Theilhaber expanded the scope of these investigations, while still applying statistics and sociology to medicine, in *Der Untergang der deutschen Juden.*

With Theilhaber and the other Jewish scientists discussed in this chapter, we come to a point of methodological symmetry. If Joseph Jacobs's anthropology was dominated by sociological analysis and Samuel Weissenberg's by numerical analysis, the Zionists can be said to have successfully merged the two approaches and used statistics to bolster their sociology. The Zionist race scientists attempted to bring to their political agendas a scientific detachment and body of evidence that they believed could not be dismissed as mere propaganda. They presented their critiques of Jewish life in the Diaspora and made their claims for the establishment of a Jewish national homeland on the basis of statistical proofs they held to be unassailable.

Both Theilhaber and Ignaz Zollschan, whose work will be treated shortly, were derided in the popular Jewish press as hawkers of what was called the "literature of dissolution and disintegration" (*Auflösungsliteratur*).[49] Taking as its central point the inevitable disintegration of first German, then Western European, and finally all Diaspora Jewry, authors of such texts painted a bleak picture of the alarmingly rapid social decay of Central European Jews.[50] Theilhaber maintained that the Jews of Germany were confronted with the inexorability of their own demise due to "low fertility, conversion, intermarriage, the increase in celibacy, venereal diseases, mental illnesses, the movement from the country to the big cities, and the entrance into free and commercial professions at the expense of artisanal work." "All these factors," he wrote, "signify an extreme danger to our existence."[51]

Theilhaber recommended that drastic measures be taken to stem the tide of Jewish demographic loss. He declared that only a unified effort could serve the needs of fast-disappearing German Jewry, and to this end he insisted that the individual Jewish communities, the *Gemeinde,* be nationalized. Only then would they be in a position to encourage an increase in the fertility of Jewish families by instituting a rational birth policy (*Geburtenpolitik*). This, he argued, should also include a program of communally funded, pecuniary incentives and subsidies, as well as a eugenics program.[52] For Theilhaber, as a Zionist, all such prophylactic measures were to be a means to an end, and not the end in itself. It was only Palestine, that "lustrous reservoir of Jewry," that offered any demographic hope for the future of the Jewish people.

Theilhaber's *Der Untergang* met with a volley of criticism from all corners of Jewish institutional life in Germany. The prophesy that this ancient community, which had withstood the tribulations imposed by so many enemies in its long past, would die out by its own doing was unimaginable. Often deeply divided, German Jews, Orthodox, Liberals, assimilationists, and even some Zionists were firmly united in their wholesale rejection of what came to be known as Theilhaber's *Untergangstheorie*. Yet despite the visible unanimity of opinion in rejecting Theilhaber's prediction, slight nuances and shades of emphasis point to acute differences within the Jewish community concerning its future in Germany. Furthermore, the criticisms that Theilhaber faced represented the forces opposed to Zionism. An examination of the debate over the *Untergangstheorie* therefore reveals the inner Jewish dynamics that spurred the Jewish race scientist's enterprise. Although antisemitism always loomed menacingly in the background, the Zionist scientists in Germany, in accordance with the Zionist philosophy of Jewish self-determination, regarded the contemporary social realities of German Jewry as the greater threat to the community's well-being.

While agreeing with Theilhaber that the German Jewish community faced an uncertain future, the dominant newspaper of the Orthodox community, *Der Israelit*, rejected his analysis, especially that part which identified the movement to the big cities and the eager participation in the capitalist system as being the root causes of impending Jewish doom. For *Der Israelit*, both the causes and the solution were much simpler: "We have asserted and assert that the dismal situation of West European Jewry is exclusively brought about by the alienation from historical Judaism, that is, Torah and Mitzvot.... German Jewry, and this is the key to what we draw from the numbers, will, despite birth policies and nationalism, disappear with deathly certainty if it does not turn back to the Torah."[53] Certain that God would not forsake the Jews, another paper expressing the interests of the Orthodox community, *Die Laubhütte*, stated that "Theilhaber's book correctly points to the dangers facing modern Jewry. [But] the title *Untergang* is, for a believing Jew, unacceptable because it contradicts the Divine promise... 'I will not let you perish'" (23).

Liberal Jews, whose views were represented in the *Allgemeine Zeitung des Judentums*, read that it was a fallacy to reduce the Jewish community to so many statistics, because that which provides it with its eternalness, its ethical superiority, cannot be numerically measured (23–24). Another Liberal journal, the *Israelitische Wochenschrift*, was more charitable in its review of *Der Untergang*. It noted the danger signs facing German Jewry, and even went so far as to claim that they not only marked the road upon which German Jews traveled but also indicated that the paths to be taken by other communities could be equally calamitous. As one of the editors, Rabbi Dr. A. Tänzer, noted, "The danger of demise is a general Jewish question." As far as Germany was concerned, "the surplus of births in recent years is in no way sufficient to cover the loss through conversion, intermarriage, and emigration, etc., and without the immigration [of East European Jews] one could already record a visible deficit among German Jews." But for Tänzer, no elaborate plan of action such as that suggested by Theilhaber, and certainly not emigration to Palestine, was required; the debilitating consequences of conversion and intermarriage could be halted if alienated Jews were reintroduced to the fundamental principles of the religion. Approvingly, he wrote that "the modern Liberal-Jewish movement has taken upon itself the task of winning back the estranged Jews to Judaism, without the absolute obligation to [follow] particular forms.... The preservation of Judaism as a religion, and not the Jews as a race, is the life's duty of the Jewish community."[54]

The most damning criticism of Theilhaber's book was in *Im deutschen Reich,* the official publication of the Centralverein deutscher Staatsbürger jüdischen Glaubens, the largest representative body of German Jewry. In a blistering fifteen-page review of *Der Untergang,* the Berlin statistician and physician Jacob Segall challenged Theilhaber's scientific method, accusing him of being generally careless with the statistics (many of which he claimed were incorrect) and too rash in drawing conclusions from them.[55]

In a surprising act of communal solidarity, even the main organ of the German Zionist movement, *Die Welt,* put aside its bitter enmity towards *Im deutschen Reich* to concur with Segall, whose rather circumspect statistical analysis had led him to a more optimistic prognosis of the future of German Jewry. *Die Welt*'s reviewer noted that "a return to Jewish laws, and likewise the nationalization [of the Jews], will not occur in the near future, and for that reason one would have to doubt the sheer continuance of German Jewry were Theilhaber's data faultless. But out of love for his people, Theilhaber has become a pessimist [*Schwarzseher*]. . . . According to Segall, an exact fertility statistic is today impossible to ascertain, for the preliminary scientific work is lacking" (24). Another important Zionist periodical, *Ost und West,* however, gave Theilhaber's book a slightly more positive reception. The reviewer wrote: "I am not enough of a specialist to be able to judge, but I have the impression that even if not everything is correct with his numbers, they nevertheless speak a persuasive language. And in the end, one doesn't need statistics to clearly see that something is rotten in the state of Denmark."[56]

Theilhaber had hit directly upon the collective raw nerve of Germany's Jews, especially their official representatives. He apportioned a great part of the responsibility for the current social status of the community among the leaders themselves, declaring that "the Jewish leadership has, in its frivolous handling of the question, made itself really comfortable by avoiding the question of the demise of the German Jews" (21).

The communal leadership, insulted at Theilhaber's rebuke, struck back. At a lecture before the governing body of the Berlin Jewish community, the privy councillor, Dr. Stern, declared with pride:

And if in this hour I propose the question: Will German Jewry disintegrate? I, from a layman's standpoint, unreservedly disavow the question. I triply oppose this burning question when I gaze around these rows of tables with admiration, joy, pride, and satisfaction at the gathering of hundreds of men from the most distinguished and prominent professions and positions in life—luminaries of science, leading artists, phoenixes of commerce, pathfinders of industry, pioneers of technology—and identify them as the appointed representatives of German Jews and their interests. Face to face with this highest, most delightful scene, I cannot refrain from quoting the words of the psalmist, 'the genius of Israel neither rests nor slumbers.' (22)

Despite the fact that many disagreed with both Theilhaber's methodology and prescriptions, very few, perhaps with the exception of Segall, the Centralverein, and the self-satisfied privy councillor Stern, challenged his basic claim that German Jewry was demographically weak. Theilhaber's harsh critique and the solutions he offered were in keeping with the approach of all those who attempted to map out the Jewish people's biological past and future. He, like the other Jewish race scientists, drew from the well of historical knowledge in order to demonstrate just what social problems he believed confronted the community at the time.[57]

Theilhaber could perhaps be classed as one of the earliest historical demographers, for he was concerned with demonstrating the demographic consequences of certain patterns of social behavior. To this end, he pondered the historical reasons for the fall of ancient empires. In his view, ancient Rome had collapsed because of the absence of a population policy. In other words, insufficient attention had been paid to the problems of fertility and the maintenance of self-regenerating population levels. This, of course, immediately begged the question: What was the key to the perseverance of the only *Kulturvolk* of antiquity, namely, the Jews? For Theilhaber, "it appeared to be bound up with their sexual-hygiene institutions and customs" (7). According to this view, the story of Judaism's birth as elaborated in the Bible was neither pure fantasy, as was the epic of Gilgamesh, nor a masterful mix of history and myth, like the *Iliad*. Rather, the purpose of religious proscriptions in the Bible was to encourage individual discipline in order to create a people whose very basis was anchored in moral and ethical thoughts and aspirations. Surrounded by foreign nations and constantly threatened by assimilation, the Jews developed a sexual ethic whose primary goal was intracommunal, healthy, and fertile reproduction. For the biblical prophets who encouraged this, it was a moral act, and indeed it became what Theilhaber called the "categoric imperative of Judaism." Fertility in the service of God and the people became the highest goal. "The history of the Jewish people," he wrote, "the wondrous preservation of its existence, cannot be conceived without these sexual arrangements" (9–10). With these words, Theilhaber made explicit the connection between Jewish biological and historical destiny, a link acknowledged by all race scientists.

In fact, it was not only race scientists who held this view. In "The Correspondence of an English Lady on Judaism and Semitism," the great nineteenth-century Jewish historian Heinrich Graetz noted that through the Ten Commandments "Judaism reveals clearly and unmistakably the simple though seldom accepted insight that the survival or demise of nations is determined by the observance or violation of the laws of sexual morality." Theilhaber was merely echoing these words of Graetz: "Can one fail to appreciate, even for a moment, that the fundamental law of Judaism sets up sexual restraint as the elixir of life not only for the individual but for the entire national body to insure the survival of the community?"[58] So by the time Theilhaber expressed his own similar ideas, the view that there was a link between the sexual practices of the Jewish people and their longevity as a historical community was well established. Nevertheless, Theilhaber was prompted to make such claims for different reasons than Graetz, whose musings are more blatantly apologetic.

Theilhaber's emphasis on sexual hygiene must be seen in the context of contemporary developments in German and, more broadly, European medicine at the time. Ideas of racial fitness and improvement, eugenics, and the promotion of healthy sexual practices that were of societal advantage and in keeping with bourgeois notions of morality dominated the discourse of social reformers and medical practitioners. Those taking part in the debates often cited Jews as the consummate bearers of a sound eugenic tradition and a code of healthy sexual hygiene practices. Jewish doctors such as Theilhaber stressed and even publicized this notion. For example, the health and sexual practices of the Jews was the chosen theme of the Jewish exhibit at the International Hygiene exhibition held at Dresden in 1911.[59]

Theilhaber, like Elias Auerbach and the other Zionist doctors, painted an idealized picture of Jewish medieval life, one that saw Judaism's reproductive "categoric imperative" fulfilled. Of central importance for Theilhaber was the fact that up until the eighteenth century, Jews made no distinction between their religion and their nationality. The one was inextricably bound up with the other. Thus, the biblical entreaty to go forth and multiply was heeded on the individual level as well as promoted for the demographic and sociological benefit it would bring to the nation as a whole.

But at the end of the eighteenth century, a transformation took place within Judaism in the wake of the French revolution and the subsequent realignment of society and the individual's and community's relations to it. From this time on, "the Jewish nationality, with its pronounced *völkisch* religious system of government [*Religionsverfassung*] transformed itself into a religious community [*Religionsgemeinschaft*]" (41). Gone was the national "categoric imperative" to be fertile for the benefit of the race. From that time on, Theilhaber said, the disintegrative characteristics of assimilation began to take their toll on Jews as they abandoned their ancient national-religious customs. With the bonds of tradition loosened, the Jews of Germany saw, over the relatively short span of one hundred years, their community decimated by infertility, intermarriage, and conversion.

In numerical terms, the percentage of Jews as part of the whole population of Germany was on the decline. Although the number of German Jews increased from 512,153 in 1871 to 615,021 in 1910, the rate of increase of the Jewish population in each of these four decades was far outstripped by that of the general population.[60] This is clearly illustrated by the dramatic decline in the ratio of Jewish to non-Jewish births. In Bavaria, for example, in 1876 there were 34.40 Jewish births per thousand Jews; by 1906 the figure had dropped to 18.47.[61] The corresponding figures for the Christian population were 45.90 and 35.91 per thousand. In the north, Prussia showed an even greater decline in the Jewish birthrate. Statistics that were taken over an eighty-six-year period, from 1822 to 1908, show that the rates dropped from 35.46 per thousand in the period 1822–1840 to 23.75 in the period 1888–1892 and finally to 17.01 in 1908.[62] Indeed, in the decade of the 1880s, there was an absolute increase of only 6,272 individuals, a mere 1.12 percent. Three developments largely account for this situation, all of which were of major concern to the Zionists: emigration to the United States, the increase in conversions from Judaism to Christianity, and the rising number of intermarriages. Theilhaber correctly assessed that the increase in the Jewish population of some 40,000 to 50,000 people between 1890 and 1910 was primarily due to the mass immigration of East European Jews. In other words, the indigenous Jewish population of Germany could not be relied upon to produce enough individuals to ensure the community's continued growth (47).

Furthermore, not only was there a decline in the percentage of Jews in the population, but in twelve German states there was an absolute decrease in the number of Jews. Alsace-Lorraine's 39,278 Jews in 1881 equalled 2.5 percent of the state's total population, but by 1911 the figure had sunk to 30,483, or 1.6 percent. The Jewish population of Hesse had also decreased in both real and percentage terms, as had those of Baden, Württemberg, and Mecklenburg-Schwerin. In 1840 Bavaria's Jewish population was 59,168, but by 1910 it had decreased to 55,065 and no doubt would have sunk much further if not for large-scale East European Jewish migration into the area. The most telling statistics were from Prussia, where despite a small but steady increase in the absolute number of Jews from 1871, when they numbered 325,426, or 1.32 percent of the total population, their percentage of the total population had decreased significantly by 1910, when 415,867 Jews made up only 1.07 percent of the total population (53).

For Theilhaber, the low birthrate and high number of conversions and intermarriages were in part a function of the mass migration of Jews to the cities from rural areas. In the most unflattering terms, he declared that "in the Jewish character [*Volkscharakter*] lies the striving after sociability. Just as a butterfly flutters around a light, so the Jew, with his sensitive, nervous character searches for the allurement of the big city. He is enticed by the sights of material and intellectual gratification" (69). Thus, Jewish assimilation to the ways of the city and non-Jewish customs, as well as their transformed occupational structure, spelled demographic and sociological disaster.

This image of the restive, uprooted, urban Jew, so common to much of the antisemitic literature of the early twentieth century, was also a mode of discourse common to much German Jewish literature, be it Zionist or assimilationist. We see it in the Zionist Max Nordau's call for the creation of a *Muskeljudentum* (1898), in the assimilationist statesman and industrialist Walter Rathenau's exhortation to Jews to "dedicate a few generations to the renewal of your outer appearance" (1897), and in many other Jewish writings on the Jewish question.[63]

Another standard feature of this type of literature was its anticapitalism. Theilhaber expressed his antagonism toward the capitalist system, charging that it was chiefly responsible for the dangerous decline in the Jewish marriage rate and its most obvious consequence, the drop in the Jewish birthrate. Referring approvingly to Werner Sombart, Theilhaber pointed to the dominating role played by Jews in the development of the free-enterprise system and to what he called the "Judaizing" of the big city. From at least the second half of the nineteenth century, he said, we begin to see the prominent Judaization [*Verjudung*] of particular branches of industry. Cattle dealing was an old Jewish domain, as was merchandizing. The Jews have long dominated the grain trade and likewise the leather industry. [But] there developed in [the 1880s] industrial undertakings which made shoe manufacture, confectionery, and chemicals great enterprises in Germany. Jews dominate the stock exchange and banking, and monopolize the legal profession and, in part, the press, medicine in the big cities, theater, publishing, and much more. This time of discovery always sought to open up new enclaves. Jewish influence Americanized commerce (the department store), mercantalized branches of industry (AEG-Hapag), and created modern advertising (Mosse). (46)

Despite the tone of suppressed pride in Jewish commercial achievements, the "time of the conglomeration of capital" had become a very difficult one for the Jews. It was the gentile proletarian who Theilhaber truly envied, because his detachment from the entrepreneurial side of modern capitalism afforded him a simple life, one where family and procreation were prime concerns: "The [non-Jewish] worker marries on average rather early. By his early twenties he reaches his maximum earning power. He rents out his labor, his strength, and his poverty to the employer and earns with it, through pure physical work, his livelihood. The proletarian also comes of age at the time that his sexual maturity and highest earning capacity coincide, thus providing the foundation for marriage" (74–75).

This was not the case with the Jews. Because they were most often the representatives of capital and not labor, both their spiritual and financial reserves were taken up with "the dance around the golden calf." And so while "the simple worker builds his life on his current economic situation, the Jewish commercial employee calculates the probability or improbability of the quantity of his [future riches]." Delaying marriage until financial security had come the couple's way, the Jews married later than non-Jews, and sometimes not at all. Theilhaber noted that "of one hundred Jews, twenty-five do not marry. A further quarter remain childless, another quarter restrict themselves to one or two children, and only among the last quarter do we meet those with three and, among a small percentage, more children." He showed that from 1910 to 1913, among Christians in Berlin between the ages of forty and fifty, 10.7 percent of men and 14.7 percent of women were unmarried, while the corresponding figures for Jews were 22 percent and 27 percent respectively. The consequence was a rapidly declining birth rate and, eventually, communal demise (74–78).

But Theilhaber was not a do-nothing pessimist who only lamented the communal disintegration he observed. With his characteristic practicality and optimism, as exemplified by his early political and organizational experience, he offered concrete proposals to rectify the situation. He suggested that an effective social policy be instituted to stem the tide of celibacy, late marriage, childless or one-child marriages, intermarriages, and divorce. Writing in 1929, when the financial condition of the large majority of German Jews had changed drastically for the worse, Theilhaber was convinced that poverty was at the root of these problems. Lack of fecundity was not a racial characteristic but rather a function of the desperate financial state of a community that had been badly hurt during the hyperinflation of 1923.[64] Theilhaber suggested that the Jewish Gemeinde adopt measures to improve the lives of its impoverished members. He maintained not only that they should establish Jewish hospitals to care for their sick, but that they also should build affordable housing for those currently living in slums or forced into prostitution.[65] He held that with decent and inexpensive housing, those presently prevented from marrying young would be able to do so. In part, owing to Theilhaber's lobbying, the proposals were put into effect.

Theilhaber also recommended that "motherhood be recognized as an achievement," and to this end he proposed that the community reward those mothers with an annual stipend of 500 to 750 reichmarks for each child above the minimum of two.[66] In *Das sterile Berlin* (1913) Theilhaber had already proposed a corollary to the reward system he now advocated. He suggested that families with few or no children, "child-poor" as he called them, be taxed to support those who were "child-rich." [67] Theilhaber's proposals for communal reform were comprehensive: the establishment of Jewish public welfare institutions, employment relief services, and youth centers, a complete restructuring of the Jewish education system which he declared had become a farce, the recognition by the Gemeinde of equality for children born both in and out of wedlock, and the granting of equal status within the Gemeinde to the recently arrived East European Jews. This last point was considered crucial by Zionists, for they felt that enfranchising recently arrived Jewish immigrants from Eastern Europe would tip the scales in their favor during communal elections, thus allowing for an even more ambitious project of Jewish social engineering.[68]

It should be clear by now that Theilhaber's conception of the Jewish people was not a racial one, although he used the language and methodology of race science quite freely.⁶⁹ Rather, he saw the Jews as a unique religious-national community. Tragically, that quality had dissolved in the relatively short span of two hundred years as a consequence of emancipation and subsequent assimilation. The demand emancipation made on Jews, that they cease to affirm their national character—that is, cease to be Jews—meant for Theilhaber that in the West they were merely left with the outward trappings of their religion. But since the majority of these Jews had stopped practicing their religion, their Jewish identity had all but vanished, having been replaced with the ideology of assimilation. This brought with it all those disintegrative features particular to modern Jewish life, especially as they appeared to Theilhaber in his native Germany: intermarriage, conversion, and declining birthrates. For Theilhaber, only Zionism and the renaissance of the Jewish people in Palestine would restore to wholeness the fragmented character of the modern Jew.⁷⁰ Above all, Zionism would return to Jews their lost religious-national character, one that Theilhaber regarded as the only prophylactic against what he had statistically shown to be the inevitability of Jewish "race suicide."

Ignaz Zollschan

The clearest and most unambiguous expression of the purpose of Jewish race science was made by the Austrian physician Ignaz Zollschan. As the world was preparing for war in 1914, Zollschan summed up the connection between antisemitism and the role Jewish race scientists had to play in combatting it: "Anyone familiar with modern racial antisemitism and with the latest literature will recognize the reality of these disgraceful attacks, and will understand that should such theories be allowed to remain unanswered they would become a great political danger. It is very desirable, therefore, that we should employ the same weapons as our opponents—that is to say, the weapons of anthropology, sociology, and natural science—to investigate the social value of the Jews."⁷¹

Zollschan dedicated himself to this crusade with unflagging energy. Like the other Zionist doctors, he too was driven to pursue race science because of the defection of German Jews from the community through radical assimilation, intermarriage, and conversion. But of even greater significance for him was antisemitism. This remained his primary target from the beginning of his career until the very end, and his preoccupation with it stems directly from the Viennese environment in which he spent his working life.

Zollschan was born in Erlach, Lower Austria, in 1877 and studied medicine at the University of Vienna, where, in partial response to its virulent antisemitism, he was involved in student politics as a member of the Unitas student union.[72] Upon graduation, he served for a while as a ship's doctor, then returned to Vienna, where he served as a resident at the city's general hospital. Soon thereafter he left the Habsburg capital for Carlsbad in Bohemia, where he remained in private practice until 1933, specializing in X-ray studies of the abdominal organs, about which he eventually published several monographs.[73] However, it was as an anthropologist, and later as a vocal opponent of the Nazis, that he achieved his greatest recognition.

Zollschan was a rather prolific writer, turning out many books and articles on the Jewish Racial Question, Zionism, Diaspora nationalism, and antisemitism. He was even responsible for the development of a new current within German Zionism called Binyan ha-aretz (Building of the Land), which regarded as its primary task the immediate establishment of Jewish towns, farms, and businesses in Palestine, as opposed to the propagation of Zionist activities within the framework of Jewish Gemeinde politics, as had been advocated by Theilhaber, Sandler, and other Zionists. But after World War I, Zollschan retreated from Zionism. Deeply disillusioned by the consequences of nationalism run amok in the war, he headed a movement known by its acronym, IGUL, the Zionist Fraternities of Austrian Universities, which was dedicated to divesting Zionism of its nationalism. He was unable to countenance a movement which he declared took antisemitism for granted and "reject[ed] the defense struggle [against antisemitism] as supposedly futile and senseless."[74] The Nazi threat eventually forced Zollschan to leave Czechoslovakia for London, where he died in 1948.

Even more than the Berlin Zionist anthropologists (such as Auerbach and Theilhaber), Zollschan's work was clearly impelled by the intensely antisemitic atmosphere in which he lived.[75] As the historian Robert Wistrich has correctly asserted, "The multi-ethnic Habsburg Empire was the cradle of the most successful modern political movement based on antisemitism to emerge anywhere in nineteenth-century Europe. In spite of parallels that can be drawn with events in neighboring Germany and Russia, the main components of Austrian antisemitism, its multinational character, agitational techniques, and mass impact, were distinctly novel."[76] In particular, pre-war Vienna, where Hitler later professed to having "studied" the Jewish problem, was a seat of widespread popular and organized antisemitism. The city was home to a wide variety of antisemitic forms, including the traditional Catholic antisemitism of the social reformer Karl von Vogelsang, the occultist racialism of Jörg Lanz von Liebenfels, and the Austrian pan-Germanism of Georg Ritter von Schönerer.

Zollschan's contribution to the Jewish struggle against the antisemitic movement was his 1910 book, *Das Rassenproblem unter besonderer Berücksichtigung der theoretischen Grundlagen der jüdischen Rassenfrage* (The race problem especially considered on the theoretical basis of the Jewish racial question).[77] Over the next fifteen years the book went through five editions, assuring Zollschan a prominent position among both Jewish and gentile anthropologists who worked on Jewish racial problems. That his apologia was written primarily out of concern for the Jewish future in this hostile climate is clear from the introduction, where he wrote despondently that "at the present time, the process of the destruction of the Jewish race is underway." Despite the difficult atmosphere for Jews in Vienna, 1910 was rather early for such desperate prognostications, and like his contemporary, Felix Theilhaber, Zollschan was regarded in many quarters of the Jewish establishment as yet another *Auflösungsapostel*.

It has been claimed that *Das Rassenproblem*, Zollschan's most important work, was a refutation of the Anglo-German racial antisemite Houston Stewart Chamberlain and his book *Foundations of the Nineteenth Century*.[78] Yet this is inadequate as a full explanation for Zollschan's response. Chamberlain's book appeared in 1899, and Zollschan's reply would therefore appear to be rather late in coming. Although he was in Vienna from 1889 to 1909, Chamberlain, who married one of Richard Wagner's daughters after the composer's death, was also part of the circle at Bayreuth ruled by the widow Cosima Wagner. Chamberlain consistently identified himself with Germany and Wagnerian Germanic ideology, not the multinational Habsburg empire.[79] Moreover, Zollschan, like Theodor Herzl, who was the Paris correspondent for the *Neue Freie Presse* at the time of the Dreyfus affair, was not drawn to the Jewish problem by one single event or person; and like Herzl, Zollschan did not have to look outside Vienna to find an antisemite to refute. In fact, from the 1930s on, he mounted a one-man crusade against the race theories of the Nazis, that is, long after Chamberlain's death.

It makes more sense to view *Das Rassenproblem* as a response to what Zollschan both saw and felt as a young Jewish medical student in the city. He bore personal witness to Karl Lueger's reign as the antisemitic mayor of Vienna, which by coincidence ended in 1910, the year *Das Rassenproblem* appeared. Moreover, his student days were spent at a university that was a hotbed of anti-Jewish activity.[80] By 1885, Jews constituted 41.4 percent of the students in the medical faculty. Intense competition there and in the law faculty, where Jews were 23 percent of the student body, exacerbated the situation in an already antisemitic city. After the publication of Theodor Billroth's polemic against Jewish medical students in 1875, anti-Jewish riots erupted at the university, and frequently thereafter antisemitic students broke up lectures with cries of *"Juden hinaus."* By 1890, all the *Burschenschaften,* or duelling fraternities, were antisemitic. The low point came with the decision of the conference of the duelling fraternities at Waidhofen in 1896, when they declared their refusal to give satisfaction to Jews because they "lack honor or character."[81]

Fin-de-siècle Vienna was not only a city with an overtly antisemitic political climate, but was also the home of a vibrant artistic and scientific culture. Viennese Jewry played a conspicuous role in this local renaissance, providing an inordinately large number of artists, writers, musicians, and scientists, especially doctors.[82] Into this highly charged atmosphere came masses of East European Jews. From 1880 to 1910, Vienna's Jewish population dramatically increased from 72,543 to 175,318, or 8.6 percent of Vienna's population.[83] The presence of this "foreign" element at a vulnerable, indeed electric time in the city's history helped heighten the anti-Jewish backlash that culminated in the election of the antisemitic mayor, Karl Lueger, in 1895. It also sparked great tension between an insecure Germanized Jewish elite and the Yiddishized Galician masses.[84]

Zollschan's *Das Rassenproblem* must be seen, then, as a response not to a single individual but to an entire constellation of events. It must be viewed against the backdrop of antisemitism, advanced Viennese-Jewish assimilation, East European Jewish immigration, the Jewish role in the advance of modernism in culture, and the establishment of political Zionism by the Viennese Herzl. To the antisemites, Zollschan stressed the cultural and historic value of the Jewish people, emphasizing their contributions to humanity. In response to Jewish factionalism, he maintained that the Jews formed a single, homogeneous racial type. To the assimilationist Jews, his anthropology contained sharp strictures against those who were misguided enough to think that cultural assimilation altered the biological facts of race. Because of this, and the all-pervasive antisemitism of European society, Zollschan maintained that only Zionism, with its promise of Jewish national rebirth in Palestine, would solve the Jewish racial question.

Lueger is supposed to have said, "I decide who is a Jew," but it was Zollschan who tackled the stickier question: What is a Jew? *Das Rassenproblem* is divided into three separate sections. The first is an elaboration of race science in general and the various opinions concerning the racial qualities of the Jewish people. The second examined the historical factors concerning racial mixing and the relative importance of environmental influences on the development of the psychology of races, the Jews in particular. The third and final section of Zollschan's book, a study of the cultural worth and legacy of the Jewish "race," is the most important. Zollschan said as much when he noted that because physical anthropology was still in a "primitive state," nearly all researchers based their findings on historical, as opposed to scientific or anthropometric, arguments. For Zollschan, the crux of the Jewish racial problem lay in determining the intellectual and spiritual value of the race. This was so because his target, the modern antisemitic movement, challenged the Jews on the basis of the ideology of Aryan supremacy—a belief that the Jews were culturally and intellectually inferior solely because of their Semitic origins.[85] Zollschan's answer to the question, What is a Jew? was based on his conviction that the Jew was a member of a *Kulturvolk*. Jews were the bearers of an intellectual and spiritual tradition at least as great as that of the Germans, and Jewish civilization, by Zollschan's account, had introduced the concept of ethical justice to the rest of humanity.

Zollschan's fundamental intent in the book was to refute the Indo-Germanic theory of racial superiority. His argument hinged on three points: that no link existed between race and language, hence to speak of an Indo-European or Semitic race was patently absurd; that the existence of tall, blond, blue-eyed Jews was not the result of intermixture with Aryans, as had been suggested in some Jewish scientific circles; and that the notion that all great achievements—artistic, social, and political— were the doings of long-headed, blond, blue-eyed northerners was both chauvinistic and unscientific. Like Aron Sandler, he warned against the hasty substitution of one exclusivist ideology for another, declaring that his aim was not to assert the superiority of Jews over non-Jews but merely to point out their equality by building a case for their racial uniqueness and integrity. In fact, striving for objectivity, Zollschan conceded that the Jews as a group have many unrefined characteristics and certainly some unwanted pathologies. He simply intended to demonstrate that these were not conditioned by heredity but by environment (4).

Zollschan premised his racial anthropology of the Jews on the belief that they formed a single racial type. He dismissed the notion that there were any racial differences between Sephardim and Ashkenazim, asserting that Jews, irrespective of geographical location, resembled one another:

> He who has attained the ability to recognize the anthropological type beneath the social one will no longer be misled by apparel, style of beard, external forms, or an artificially inculcated lively or cool temperament. Similarly, he who can disregard the various national costumes will be extremely surprised when he recognizes completely identical types among Persian Jews and Moroccan Jews, when he recognizes in the synagogues of old-established Arab Jews in, say, Cairo or Beirut, not only exactly the same faces as in East Europe, but also the same chaotic variety of Andree's "fine" and "plump" types, long, medium and broad heads, light and dark pigmentation, the same contrast between red hair and blue eyes on the one hand and the deepest black on the other. I saw the same alternation in the old Jewish streets of Haifa, among the Sephardi Jews of Constantinople, Corfu, Amsterdam, in the ghetto of Venice, among the Jews of Rome, as among the Russian and Rumanian Jewish immigrants in America, and among our familiar Hungarian, Bohemian, and German Jews.

All over the world Zollschan met with "identical types" among Jews, which to his thinking proved their "racial homogeneity." He could only conclude that "the present-day Jews . . . constitute a type which is uniform to a high degree, irrespective of the geographical terrain and the racial characteristics of the natives."[86]

Zollschan's anthropological description of the Jews as a "uniform, self-contained race" was a blatant polemic. He contemptuously dismissed all outward measures to blend in with surrounding populations, such as the adoption of indigenous forms of dress, coiffure, and deportment, as futile and largely degrading. This anti-assimilationist strain was central to the anthropology of the Zionist race scientists.

For Zollschan, the Jews retained their Jewish characteristics and appeared phenotypically different from the peoples among whom they lived because they had, as scores of scientists had previously asserted, remained isolated for two thousand years, avoiding any intermixture with native populations. He ruled out intermarriage and large-scale conversion, maintaining that religious proscriptions and ghetto isolation had preserved the Jews from contact with foreigners. In this respect his views were identical to those of Elias Auerbach. If Jews did appear to possess certain physical characteristics, such as small chests, weak musculature, pale coloration, and a melancholic look, these features were the "sad heritage from the time of dispersal." But they were merely secondary characteristics stamped onto a pre-existing form (44–45). They were not innate racial traits, and their disappearance was to be expected as soon as Jewish living conditions improved, that is, when exile came to an end. For example, he noted that "to a great degree, the size of the body lies under the influence of the environment. It is a plastic factor, and as a racial characteristic, is conditioned by nourishment, housing, occupation, illnesses and moral factors, etc." (73).

To make this case, Zollschan pointed to two specific areas where environmental influences had left an imprint on the Jews. The first was in the area of international sport, where, according to him, their recent achievements, after only one generation's release from the ghetto, testified to the value of healthy living conditions. The other was the benefit that accrued from Jewish ritual laws and sexual-hygiene practices. One of the chief boons of such customs was the absence of alcoholism among Jews. These two areas of Jewish success, both derived from environmental conditions, made for the Jews' physical capacity for regeneration (*Regenerationsfähigkeit*). As Zollschan put it, "despite the poverty of the Jews today, they can be seen as decadent, but not as degenerate" (78). At a time in European culture when races, and certain of society's "outsiders" such as criminals and prostitutes, were often categorized as degenerate, Zollschan's stress on the regenerative capacity of the Jewish people struck a defiant chord in his polemic against the proponents of the Aryan theory.

Like all race scientists, especially those who subscribed to the theory of Jewish racial homogeneity, Zollschan had to account for the apparent physical differences between various Jewish populations. A confirmed Darwinist, he believed that physical differences among Jews were caused by natural and constant deviations among individuals and groups. Deviations also resulted from the ever-changing conditions of the environment. These in turn were further developed and passed on, while older forms were sometimes completely lost. As far as Jews were concerned, this theory would explain the existence of blond Jews in northern climes and brunets in southern ones (66–73).

Zollschan still had to account for the existence of both dolichocephalic and brachycephalic skull indexes without posing any challenge to his theory of Jewish biological isolation. He wholly rejected that part of the Aryan theory which proposed that in those lands now populated by dolichocephalic peoples, the autochthanous, brachycephalic peoples had been either killed, driven off, or assimilated into the ranks of the conquering long-headed peoples.[87] Rather, he subscribed to the idea, originally proposed by Joseph Jacobs in 1899, that the skull was not the unchanging racial characteristic it was widely thought to be, and that environmental conditions, namely, a largely intellectual existence, had led to increased brain capacity in Jews, which in turn resulted in a rounder and larger cranial cavity.[88]

Clearly, a belief in racial purity did not exclude a belief in the power of the environment to change bodily form. In fact, Jacobs and many of the Zionist anthropologists amply demonstrated how it was possible to create an anthropology of resistance without betraying their allegiances to Enlightenment values concerning human adaptability. The German Zionists, who celebrated the Aufklärung and the liberal ideology that flowed from it, were able to come to a unique reconciliation of the concepts of racial exclusiveness and environmental determinism. This ensured that Jewish race science was anchored in a hopeful and redemptive liberal humanism rather than the counter-Enlightenment pessimism and jingoism characteristic of non-Jewish, and especially German, race science. Zollschan recognized this himself when he wrote, in an unpublished seventeen-point thesis on the nature of racialism, that "the racial theory was supported not by the Nordics alone. The Latins, the Slavs, the Jews, all had their own racial theories, but with them it has not assumed those exclusive, monopolistic forms bound up with domination over others."[89]

If the Aryan theory of Germanic supremacy and conquest was held to account for the existence of Nordic-looking peoples among dissimilar population groups, the belief in an inheritable Germanic psychology and intellect was even more widespread. Concepts of honor, valor, loyalty, and creative genius were regarded as typical German personality traits. And just as physical characteristics were held to be permanent, so too were these mental ones. But Zollschan dissented from this view and asserted, as he had concerning bodily change, that the environment could influence the mental qualities of a people. This discussion accounted for much of the second half of *Das Rassenproblem* as well as many of his other works.

The theory of Nordicism, which pitted Aryan against Semite, was portrayed by the antisemites as a battle of the creative German genius against the aping Jewish spirit. Zollschan's defense consisted of arguing that the intellectual properties of races were not inheritable and indelible characteristics fixed for all time. That Germany had produced a string of brilliant musicians, for example, did not mean that the nation was forever endowed with unique musical ability. Likewise, if Jews had shown themselves to be successful merchants, it was because they had been historically confined to that role, and not because they were biologically predestined to trade. The conspicuous poverty of the overwhelming majority of world Jewry made this abundantly clear. Zollschan elegantly summed up his belief in the determinative influence of the environment for the intellectual development of a people:

> The productive qualities of different races are not constant. With any people at any particular epoch they depend not on the race, but on the environment at the time, plus the environment of former generations. That is the conclusion of scientific investigation into this question, of the laws of heredity and acquired character, and likewise of the results of historical inquiry. With the different stages of culture through which a people progresses, its psychology changes. It is the psychology of these stages, and not of race, which is responsible for the spiritual condition of any people.[90]

But Zollschan's theory about the development of national character was more involved than this. He did not entirely deny that race played a role in the development of the intellectual life of a people. Essentially, he differentiated between the transmission of what he called the quality of aptitude (*Begabungsqualität*), which was environmentally determined, and the quantity of aptitude (*Begabungsquote*), which was racially determined. According to Zollschan,

> The descendants of one race may indeed be more gifted than those of another. The explanation is to be found in the past experience of that stock. In the entire organic world, we find that every being developed and perfected those organs which were mostly employed. The limb which is most exercised, grows best. When it was necessary, therefore, for a certain species to develop its brain to the highest perfection—when a certain race, by its own free-will or by force of circumstances, devoted itself to work which required it to perfect the brain, it necessarily follows that the descendants of such a race have the advantage over the descendants of another race. The quality of their ability . . . depends upon the environment, the stage of development, and the influences of tradition; but the quantity of their capacity, the magnitude and intensity of their ability does not depend upon environment, but upon race, or rather upon the cultural activity of their ancestors. This is, therefore, a factor of heredity.[91]

As this applied to the Jews, and to their position vis à vis Aryans according to Aryan supremist theory, Zollschan drew predictable conclusions: "Now with what people and with what race was the cultural activity of their ancestors greater than with the Jews? For with the Jews study was a religious duty, and those among them who did not possess a high degree of intellectual activity were not fit for the struggle for existence. In consequence of the intensive cultural activity of their ancestors, the Jews must possess the maximum sum of innate ability." For Zollschan, the supposed distinction between intellectual acumen and creative ability was spurious. He countered assertions that the Jews had not contributed to the flowering of Western culture by pointing to the fruits of Semitic "creative genius" in ancient Mesopotamia and medieval Spain, to which he traced "the origin of Humanism and of the Renaissance of which Europe is so proud."[92] With this and a host of other examples of Jewish contributions to civilization, Zollschan affirmed the cultural value of the Jewish people.

Because he had declared *Begabungsquote* to be hereditary, his next aim was to quantify Jewish genius. Joseph Jacobs had undertaken a similar task more than two decades earlier in his "Comparative Distribution of Jewish Ability"; but there is a striking difference of motive in Jacobs's and Zollschan's pronouncements, one that reflects differences in British and German society and anthropology. Zollschan's work was defensive, a response to an all-pervasive antisemitism, whereas Jacobs's, read before the Anthropological Institute of Great Britain and Ireland in 1886, was an almost recklessly confident exercise in statistics, carried out using the methodology of his friend Francis Galton. Jacobs, whose results showed the Jews to be more intelligent than the English and the Scots, concluded his study in a manner unthinkable for a Jew living in antisemitic Austria: "I have had to risk the imputation of bad taste, and shall be content if I avoid that of bad science. I can only say that I have throughout been conscious of the danger of being biased in favour of Jews, and have guarded against it to the best of my power, taking a final precaution in warning the reader of the fact. At any rate, I do not think the results I have reached run counter to any common impression, and certainly not, in liberal England, to any popular prejudice."[93]

Thus, in "liberal England" the sum total of Jewish achievements could be lauded by Jew and gentile alike. In fin-de-siècle Central Europe, whose culture was in part molded by the widespread belief in an irreconcilable difference between Germanic and Jewish intellect, the antisemites attempted to establish insurmountable barriers between the two groups. Zollschan and other Zionists read the history of the last century as the futile and often disastrous attempt of Jews to scale those barriers. For them, emancipation and assimilation, the basis of the Jewish rise into the *Bildungsbürgertum,* had been an unmitigated disaster because it had served to break down Jewish identity, decimate Jewish culture, and compromise the biological purity of the race. As Zollschan put it, "[Christian] culture and civilization have brought nothing but misfortune [to the Jews]; they have estranged many of its best sons, and through political and economic antisemitism, have slowly but surely taken away the ground from under the feet of the great masses."[94]

Turning the problem around, Zollschan asked rhetorically, "Would it not be perhaps of great benefit to the development of civilization if the Jews were to assimilate with other races of high standing? . . . We must seek for an answer in Sociology, History and Natural Science. Which is better when considered from the general point of view, race-mixture or race purity?" After weighing up the evidence, Zollschan found in favor of retaining racial purity, preventing miscegenation, and above all halting the increasing practice of Jewish intermarriage. Nevertheless, the sum total of the social conditions of European Jewry made Zollschan a very pessimistic man. In 1914 he wrote dejectedly, "After thousands of years of splendid development and stubborn resistance, [the Jewish race] now presents the sad picture of the body of a people which is partly perishing in misery and partly in the course of decomposition."[95]

It is at this point that one can expect to see the confluence of Zionism and anthropology. Zionism was central to Zollschan's case, for as far as he was concerned, it provided the only hope for the salvation of the Jewish people, who were being attacked from without and were disintegrating from within.[96] Zionism had a twofold purpose in Zollschan's schema. On the one hand, it would bring home to the antisemites that Jews possessed a creative energy sufficient to reawaken their national spirit. On the other hand, by pointing with pride to the contributions the Jews had made to civilization, it would win back to the fold those who had become estranged from and ashamed of their roots, the assimilated Jews who saw nothing of value in Jewish culture and longed to be regarded as Germans. Summing up the effects of the twin historical forces of antisemitism and assimilation, Zollschan wrote bitterly, "Scientific investigations in the sphere of race-anthropology and race-biology have sought to establish the fact of the *inferiority of the Jewish race*, an intellectual, spiritual, aesthetic, and physical inferiority. To this result we owe, on the one hand, anti-Semitism . . . on the other, the lack of pride and self-respect in the Jews themselves, which tends to their utter disorganisation."[97]

The program of restoring pride and self-respect to the Jews was of decisive importance for the Zionists. Zollschan expressed the feelings of many like-minded activists who saw that the Jews faced very limited choices in devising a method to achieve this aim. If Zionism were to fail, then the consequences would be dire. Living up to his reputation in the Jewish press as an *Auflösungsapostel*, Zollschan declared bluntly, "Without Zionism, there are only two possibilities; the destruction of the race, or its physical degeneration" (494). Yet at this point in his career he was convinced that Zionism would eventually triumph, because only its redemptive qualities could stem the tide of Jewish disintegration.

In diagnosing the Jewish condition, Zollschan, a physician, regarded the Jews as a whole as he would have an individual patient. And like the German medical profession, which often saw in Jews a degenerate group laden with a variety of physical and psychological ailments, Zollschan observed that "the sociological structure of modern Jewry is unhealthy."[98] The reason for this was that it lacked its own soil. For Zollschan, a people's greatness could only be fully expressed in its own territory (422–433). Despite his assertions about the cultural value of the Jews, and his discussions of past achievers who were Jewish, he disconsolately pointed out that given their unhealthy social state, the Jews' true cultural worth remained an unfulfilled potentiality, one that would remain so under existing circumstances (425). The only solution, therefore, lay in the dissolution of the Diaspora and the establishment of a Jewish national homeland.

Zionism, in Zollschan's account, was not merely an abstract political concept that could be brought to fruition through the beneficent intercession of foreign powers and the collection of funds to support the enterprise. It was "more than the mere imitation of the national fashion of other peoples, more than the reaction to antisemitism" (428). In fact, Zionism would exist even if there were no antisemitism. Rather, Zionism's raison d'être was that it alone could prevent the decaying effects of assimilation (430). Jewish adoption of modern secular culture and political integration into the states in which they resided had atomized the Jewish people, stripping them of their solidarity and cohesion. This idea was directly related to Zollschan's belief that race mixing—in the case of the Jews, by intermarriage—led to racial chaos. Thus Zollschan's anthropology bolstered his political convictions.

Zollschan's tireless efforts to combat racism, particularly that of the Nazis, share the same basis as his race science: a profound opposition to biological determinism. In November 1933, in what became known as the *Zollschan-Aktion*, Zollschan submitted to Thomas Masaryk, the president of Czechoslovakia, a plan for a "scientific investigation of the theoretical foundations of racial philosophy."[99] The plan met with Masaryk's approval, and the Prague Academy of Sciences, which appointed a special commission to investigate the subject, published its findings in a volume entitled *The Equality of the European Races*. Continuing to agitate, Zollschan organized a petition "To the Scientists of Our Time" to rally the support of prominent Europeans to his cause. The petition bore the signatures of such luminaries as Sigmund Freud, Albert Einstein, Karl Barth, J. B. Priestley, and Thomas Mann.[100]

With momentum building through Zollschan's constant exhortations, the Prague academy sent out feelers to the major European universities to see how receptive members of the science faculties would be to participating in an international conference on race. The first inquiries were made in Vienna, and the results were positive. Not only did the university faculty approve, but so too did the Austrian president and the archbishop of Vienna, Theodor Cardinal Innitzer, who advised on the best way to approach the Vatican for support. In two private audiences with Pope Pius XI in 1934 and 1935, Zollschan was promised that the scientific facilities of the church would be put at his disposal were the international conference to take place. In addition to Prague, Vienna, and Rome, Zollschan met with positive responses in Paris, London, Stockholm, and Amsterdam.[101] By 1938, with international cooperation assured, the plan for the international conference in Prague came to an abrupt halt when the Czechoslovak republic collapsed.

That the failure of the *Zollschan-Aktion* was due to a constellation of events and not simply to the Nazi absorption of Czechoslovakia was immediately recognized by Zollschan himself. Whereas he pointedly accused a host of different people and groups of sabotaging his plans, Zollschan was particularly critical of the Jewish leadership, who he said "did not ripely understand the overwhelming importance of the task."[102] Among the Jewish leaders Zollschan criticized, the Zionists were singled out for particular attention. According to him, the Zionist tactic of not responding to antisemitism was entirely misguided: "For it was not merely a question of the defamation of the Jews and the assertion that they are an inferior race but the insinuation that Jews by the very nature of their 'blood' constitute of necessity the poison centre for spreading destruction among all other peoples. This insinuation and the supposition that Jews as a nation must be indifferent towards the State could and should have been combatted. There were possibilities of eradicating these nonsensical theories at their inception."[103]

These words appeared in 1943, when Hitler had already murdered millions of Jews. In an assessment of Zollschan's racial theories, the Israeli historian Walter Zwi Bacharach condemned Zollschan for believing that reason, given expression at international conferences, could in some way undermine Nazi race theory.[104] To be sure, Bacharach is correct. No reasoned dialogue with the Nazis would have worked, and certainly by 1943 it would have been absolutely useless. But, Bacharach's conclusion that Zollschan's race science was "naïve" is predicated on the belief that it had undergone a dramatic change by 1933. Clearly, by that time Zollschan had expunged any expression of biological determinism from his anthropology, even though all previous expressions of it were murky, with no firm line drawn between culturally and biologically inherited traits. But the continuity of his unflagging struggle against antisemitism, rather than any break with his intellectual past, should be stressed. Who could say that when Zollschan published *Das Rassenproblem* in 1910, and visited New York in the 1920s to discuss with Franz Boas the possibility of establishing a research institute devoted to refuting scientific racism, that his sincere desire to combat the dangerous and insidious theories of Aryan supremacy was a misguided expression of idealistic naivete?

The Bureau for Jewish Statistics

Although the study of the physical and cultural anthropology of the Jewish people only came to fruition in the last two decades of the nineteenth century, the seeds for such an enterprise were planted at the end of first two decades of the century. In 1822, Leopold Zunz, the leading light of the movement known as Wissenschaft des Judentums (Academic Study of Judaism), published his programmatic essay "Outline for a Future Statistics of the Jews." Zunz's use of the word *statistics* was more comprehensive than our current understanding of the term, embracing the entire sociology of the Jewish people. His thirty-nine point thesis amounted to a call to scholars to examine all aspects of contemporary Jewish life—languages, religion, customs, work, and private and communal life. Such topics, Zunz held, were to be approached regionally; the differences in Jewish habitats were to be taken into account and then incorporated into a whole and generalized picture of Jewish life, not as it was in the past, but as it is in the present as a result of that past.[105] In addition to all of this, Zunz called for studying the physical qualities of the Jewish people, including the causes and consequences of those properties. Furthermore, he was of the opinion that censuses should be taken, presumably to compare differences among diverse Jewish populations.

The work of the Jewish race scientists was inspired by Zunz. His program was heeded, and by the end of the nineteenth century the nascent disciplines of Jewish anthropology and sociology had begun to take root. More than this, the labors of the scholars of the Wissenschaft des Judentums, and those of the scientists examined here, were impelled by similar considerations. Essentially, the work of both was a response by university-trained Jewish intellectuals to emancipation, assimilation, and antisemitism. This triad of historical forces was particularly evident in Germany, where it clearly motivated the work of the Zionist race scientists.

In his *Gottesdienstliche Vorträge der Juden* (History of the Jewish Homily; 1832) Zunz suggested that the lowly status of the Jews in Germany was in some way linked to the neglected state of Jewish scholarship. He was convinced that by redressing the latter, and thereby demonstrating to European society the richness of Jewish culture, the Jews would achieve a "higher level of recognition, or emancipation." According to Zunz, if they could base themselves on the findings of this Jewish scholarship, rather than on biased and libelous Christian works for information on Jews and Judaism, German rulers and officials would be better able to make informed policy decisions about Jews that would bring about an improvement in their civic status.[106]

Zunz's goals apply to the Jewish race scientists. In 1904, eighty-two years after the publication of Zunz's "Outline," an institution was established in Berlin for the purposes of defending emancipation, stemming the tide of assimilation, and combatting antisemitism through the use of statistics. The Bureau für Statistik der Juden attempted to present the sociology of Jewish life according to the methodology proposed by Zunz and the philosophy of Wissenschaft des Judentums. It was responsible for the collection and publication in its journal, *Zeitschrift für Demographie und Statistik der Juden*, all manner of statistics pertaining to the Jewish present.

The bureau was the brainchild of the Jewish statistician Alfred Nossig, who in 1902 founded the Association for Jewish Statistics, which within one year established branches in Hamburg, Vienna, Lemberg, Warsaw, Odessa, Tomsk, and Bern.[107] In 1903 the association published its first volume of essays, edited by Nossig and entitled *Jüdische Statistik*. Arthur Ruppin, soon to be head of the Bureau for Jewish Statistics, dismissed the work as scientifically unacceptable, claiming that because the majority of articles were written by amateurs it lacked merit. He dismissed it as "beneath criticism" and, in its stead, proposed to Nossig that "we found a journal for Jewish demography and statistics in order to publish articles by experts and especially any statistical material available about the Jews throughout the world."[108]

At the time that Ruppin made his suggestion, he was not yet a full-fledged Zionist activist. He was clearly leaning in that direction, though. His book *Die Juden der Gegenwart* (The Jews of Today; 1904), with its penetrating sociological analysis of assimilation, declining birthrates, the effects of urbanization, intermarriage, baptism, and Jewish race science, came close to the Zionist critique of such phenomena.[109]

Nevertheless, despite Ruppin's embrace of Zionism shortly thereafter, the *Zeitschrift für Demographie und Statistik der Juden*, under his editorship, continued to abide by the policy that Nossig had established earlier: "The Association for Jewish Statistics is not in the service of this or that party of Jewry, but in that of the totality of Judaism and humanity, in that of science and truth. The personal convictions of the individual members of the association's leadership must never shatter this principle. . . . We grant to each contributor complete freedom in the representation of his standpoint and, at the same time, the exclusive responsibility for his statements."[110]

Thus the bureau was never truly a Zionist organization, for its membership and publications hosted a wide variety of opinions within the Jewish ideological spectrum. But articles written by Zionists did come to dominate the journal's pages. Furthermore, the whole enterprise of gathering Jewish statistics was given impetus by the Zionist movement, before which no institution for statistics had been established over the many decades since Zunz's appeal. As Nossig himself noted, "The nationally minded Jews have been . . . the most eager sponsors of the work of the Association for Jewish Statistics."[111]

Perhaps the main reason for this support was the programmatic statement issued by Max Nordau, a neurologist and leader of the World Zionist Organization, who declared at the movement's fifth congress, held in Basel in 1901, "We must know more. We must reliably learn how a people is composed [*wie das Volksmaterial beschaffen ist*], what we have to work with. We need an entire anthropological, biological, economic, and intellectual statistics of the Jewish people." This is a clear reiteration of the Zunzian program. But Nordau's address also reflects new historical circumstances, a different ideological motivation and intellectual zeitgeist:

> We must have statistical answers to the questions: How are the Jewish people physically constituted? What is their average size? What are their anatomical particularities? What are their figures for sickness and mortality? How many days are they sick on average per year? What is their lifespan? From which diseases does the Jew die? What is his marriage and childbirth

rate? How many criminals, mentally ill, deaf-mutes, cripples, blind, and epileptics do the Jewish people have? Do they have a particular criminality, and what sort is it? How many Jews live in cities and how many in the countryside? Which occupations do they engage in? How do they work and what do they own? What do they eat and drink? How do they live? How do they dress? How much of their income is spent on food, clothing, rent, and spiritual needs? All these things must be known if one really wants to know a people.

The idea behind the gathering of Jewish statistics was to effect a rational and scientifically based amelioration of the Jewish condition. For as Nordau concluded, "so long as one does not know all [these statistics], all that one wishes to do for a people will be groping in the dark, and all that one says about this people will be at best lyric and at worst empty talk."[112] Louis Leopold, the first general-secretary of the Association for Jewish Statistics, openly acknowledged the union between scientific objective and political design when he wrote that "the purpose of the association is on the one hand scientific and on the other practical, in that through its efficiency the fundamental information required for an improvement in the situation of the Jewish masses will be produced."[113] In essence, the major ideological change that had taken place since Zunz's day was the introduction of self-help or, as the Zionist leader Leo Pinsker put it, "auto-emancipation." Help was no longer to be expected to arrive from foreign rulers and legislators but would come from the Jews themselves.

Nossig's earlier claims to objectivity notwithstanding, the gathering of Jewish statistics and the writing of a Jewish sociology or anthropology based on those statistics were impelled by political considerations. Although these were often Zionist, they were also liberal and even assimilationist. These three post-emancipatory Jewish ideologies expressed widely divergent philosophies. Yet statistics gathering provided them with a common denominator in that the figures were always used to defend Jews against their detractors and to work for the improvement of Jewish conditions on the basis of those data.

Scientists representing all streams of Jewish thought were published in the *Zeitschrift*,[114] where they expounded many of the major themes of the day concerning the physical and mental anthropology of the Jews. In each issue of the journal, demographic questions were dealt with first. A section on Germany treated both national and regional issues; this was followed by sections on Europe, Africa, the United States, Asia, and Australia. Each edition featured a section entitled "Articles without Territorial Connections," which was devoted entirely to the Jewish racial question. As editor during the first three years of the journal's existence, Ruppin presided over some of the fiercest debates concerning the physical anthropology of the Jewish people. Over the course of his celebrated career, he on many occasions spoke of the Jews as a race, but his essential position was that "with regard to race, no clear line can be drawn between the Jews and non-Jews. There are no racial characteristics peculiar to the Jews only."[115]

In the discourse on Jewish physical anthropology, the *Zeitschrift* continued the traditional scientific debates of the nineteenth and early twentieth centuries concerning racial purity and its relation to intermarriage. But more than this, with its scores of articles on intermarriage and conversion the journal also sought to show in rigidly empirical terms that the defections from the Jewish community that men such as Auerbach, Theilhaber, Zollschan, and Ruppin so plaintively agitated against were as much a threat to German Jewry's continued existence as was the antisemitic movement.

As scientists, contributors to the *Zeitschrift* believed that rational, scientific inquiry could explain all sociological and anthropological phenomena. In fact, they also attempted to solve old religious problems by their scientific methodology. For example, Ruppin published a debate between Alfred Nossig and an antisemitic doctor from Munich, Curt Michaelis, on the biological significance of the Jewish idea of "chosenness." It is somewhat surprising that Ruppin permitted the vicious opinions of the Bavarian physician to be published in the journal. Yet the belief that the Jewish people professed their chosenness so as to assert their superiority over other nations was an ancient accusation, and for Ruppin its refutation was essential to establishing the scientific status of the journal. The debate also bespoke the possibilities of the scientific apologia.

Michaelis asserted that there were only two ways for a nation to assert itself—by conquering its neighbors or by promoting the self through the manufacture of inflated national pride and a sense of superiority and election. This led to the development of a racial pride (*Rassenstolz*) that could become an inherited characteristic. Among the Jews this *Rassenstolz* took the specific form of the idea of chosenness (*Auserwählungsidee*) which bore within it the concept of exclusivity, rigidity of dogma, and vanity. "The *Rassenstolz*," wrote Michaelis, "promoted race hatred in its sharpest form—the consequence of which is lasting race war. . . . The Jewish people stands principally in battle against the whole world; naturally, therefore, the whole world [is] against the Jews."[116] According to Michaelis, it was the Jews' hate-filled attitude towards the nations of the world and the reciprocal hatred of the gentiles that led the latter to incarcerate Jews in ghettos. In a twisted logic, Michaelis insisted that the *Auserwählungsidee*, itself a consequence of original Jewish hatefulness, was strengthened in the ghetto along with all other Jewish features acquired during the period of historical confinement. This produced a host of permanent and unchanging characteristics, all of them insidious.

Nossig responded by insisting that the idea of chosenness meant neither Jewish exclusiveness nor hatred of others, and that it was certainly an error to see in the application of the concept any form of *Rassenstolz*. For Nossig, chosenness implied two tasks, neither of which meant the subjugation of non-Jews: self-perfection and helping others to perfect themselves. The reward for such striving was to be the eternal existence of the Jewish people.

The goal of Israel, Nossig asserted, also brought with it a biological reward. The fulfillment of the ideal necessitated the creation of a code of dietary and marital laws whose by-products were superior Jewish longevity and reproductive powers. The holy commandment to study Torah meant the development of a profound intellectual energy and honorable code of ethical conduct toward family and others. Above all, the health of the nation was secured by the proscription against intermarriage. This ensured that the Jews never acquired those "gentile diseases" associated with excessive alcohol consumption and sexual concupiscence. Once they had retreated into the ghetto and away from the majority non-Jewish peoples, the Jews reproduced only among themselves, successfully passing on their carefully selected qualities of virtue and righteousness from one generation to the next. This process had finally reached the point where Nossig was able to crown the Jews with the title of the most "adaptable" (*anpassungsfähigste*) of all peoples.[117] The adaptability in which Nossig rejoiced was the biological mechanism that allowed the Jews to mix with the nations of the earth for the purpose of bringing them to that higher level of ethical perfection. The mission could only be achieved by the retreat into the ghetto, which permitted positive inbreeding and led to the creation of a superbly adaptable race, one capable of living in all environments, among all manner of people, and bringing to them the ethical and universally applicable teachings contained in the Torah.

This debate encapsulates not only the aims of the Jewish statisticians, sociologists, and anthropologists, but also the paradox inherent in their work. On the one hand, they sought to combat antisemitism by proving, through the statistical method, that Jews did not constitute a clear and present danger to Christian society, and that their role was one of full participation in the national agenda. On the other hand, by recourse to that method, workers at the Bureau for Jewish Statistics stressed the various qualities of Jewish life that marked Jews off from their German neighbors. These included studies that demonstrated the higher percentages of Jews attending university and the lower incidence of venereal diseases among German Jews. Jewish scientists also used the occasion to extol such positive and purportedly distinctive aspects of Jewish life as sobriety and family unity.[118]

The Bureau also addressed other themes pertinent to the Jewish racial question. For example, since Jews were charged with playing leading and often controlling roles in certain areas of the economy, researchers undertook extremely comprehensive studies to determine the true data on Jewish occupational distribution so as to counter claims of Jewish concentration in certain professions. Furthermore, impelled by the often-repeated image of the Jewish beggar, the *Schnorrer*, the Bureau's studies of Jewish welfare societies were intended to show a picture of a self-reliant, independent minority that took care of its own and was no threat to the state's coffers.

Other social ills were exhaustively investigated as part of the Jewish racial question. One of the most important of these was criminal behavior. Accusations of Jewish mendacity and dishonesty were ancient, and Judaism had of course been charged and found guilty by Christianity of the worst of offenses, deicide. By the end of the nineteenth century, the new field of criminology had begun to amass a substantial literature on comparative crime rates, and again Jews were featured prominently in the German studies. When involved in criminal activity, Jews were generally found to be most frequently guilty of nonviolent crimes such as defamation, embezzlement, the receipt of stolen property, perjury, and the forgery of documents. The question that criminology pursued was whether Jews showed a racial predisposition to commit certain crimes. In general, researchers split over the answer. Jewish students of the problem published their findings in the Bureau's journal, which not surprisingly tended to opt for an environmental explanation. The most significant point here is that Jews used sophisticated statistical studies to dispel deeply ingrained prejudices.[119] Moreover, as Nossig observed, the gathering of Jewish statistics was a prerequisite of all rational Jewish social work.

By exhaustively accumulating data in accordance with Leopold Zunz's program of 1822 and continuing the efforts first undertaken by Joseph Jacobs in 1882, the Bureau for Jewish Statistics legitimized and brought into full bloom the field of Jewish statistical sociology. The enormous quantity of studies pertaining to the physical, psychological, and sociological characteristics of the Jewish people was largely due to the efforts of German Zionism. At this unique historical moment, the discipline of statistical sociology, as defined by the work of the Bureau, was intimately bound up with the contemporary Jewish question in general and the Jewish racial question in particular.

Zionist race scientists would not, indeed could not, be too harsh on their people's past, especially its biological past. This meant perforce that they would take a decidedly environmentalist view. If a Jewish anthropologist had detected a "flaw" in the race and had been a biological determinist, then what was the point to the entire Zionist enterprise? Physical or psychological defects that existed in the Diaspora would merely accompany the Jews to the newly created Jewish national homeland. The promised salutary benefits of Zionism would remain unfulfilled, powerless to change the iron law that race is everything. So, vague claims about inherent racial instincts notwithstanding, Elias Auerbach and his fellow Zionist anthropologists were, because of their Zionism, environmental determinists.

It can of course be argued that Zionism is not incompatible with biological determinism. Jewish anthropologists could detect in the Jewish race a purity and perfection brought about by the laws of racial heredity. All that the Jews lacked was a national homeland; restoration to their ancient abode would solve what was to these anthropologists the essence of the Jewish problem: exile. To the Zionists and their anthropologists, the Jewish problem hinged on the unnaturalness of Jewish existence in the Diaspora. It was this condition of transience, living under the threat of persecution and humiliation, as well as the debilitating effects of assimilation, that were responsible for the Jew's physical and psychological state.

To Zionist race scientists, the Jewish racial problem was never to be solved by apocalyptic race wars between Aryans and Jews. They never reduced it to the crude level of the biological incompatibility of one race with another. On the contrary, the most distinguished of the Zionist anthropologists argued in their scientific texts, and in their strictly political works, for harmony and cooperation with other "races," most especially the indigenous Arab population of Palestine. Arthur Ruppin's role as a founding member of Brit Shalom, a society dedicated to Jewish-Arab cooperation and the idea of a binational state in Palestine, is eloquent testimony to this commitment.

In the light of the humanitarian tendencies of German Zionism and its understanding of Jewish history, Karl Kautsky's claim that the Zionists employed the theory of race in much the same way that "Christian-Teutonic patriots did to declare themselves demigods" simply does not bear up under the weight of evidence. Rather, it was Aron Sandler's admonition to avoid "racial chauvinism" which was heeded by the Zionist anthropologists. This did not mean that the concept of race was rejected by the Jews. It is clear that for the Zionists, race was an important category and mode of Jewish self-definition. Yet it was not presented as the sole determinant for Jewish existence. Rather, Zionist scientists wove an intricate explanatory web for the continued existence of the Jews, in which history and race were presented as working together for the benefit of the Jewish people. While it is not always certain whether race meant nation, people, culture, or ethno-religious entity, the Zionist race scientists generally accorded to the mark of history on the Jewish people the greater significance.

Furthermore, the Zionist call to awaken Jews from their historical slumber was not a rejection of the validity and dignity of that past. It is somewhat paradoxical, but the Zionist race scientists, above all because they were German Jews, were loath to shrug off the period of exile as a mere footnote to a glorious, ancient past and preamble to an even greater future. Although it is true that many German Jews, Zionists among them, were solidly, sometimes exclusively, imbued with the spirit of Germanness, Jewish history was always represented by them as a precious commodity, noble and ennobling. For them it was Jewish history that best guaranteed the continued (though threatened) existence of the Jews, an existence that they attributed to the particulars of Jewish daily life and the edifice of self-protection that the Jewish nation had established through its legal code over the millennia.

✱ SEVEN ✱

THE LIMITS OF RACIAL SELF-REPRESENTATION

The origins of modern racial thinking are to be found in the eighteenth century. Continuing encounters between Europeans and aboriginals, colonialism, and the slave trade served to diminish the avowed Enlightenment commitment to human equality. The idea that qualitative differences existed between races was a direct consequence of this meeting. Similarly, the rise of the biological sciences in the eighteenth century also gave great impetus to the process of defining human difference and in fact confirmed, with its authoritative language and methodology, popular notions of racial superiority and inferiority.

From at least that time, new biological definitions of Jewishness also began to appear. The continued perseverance of the Jews, Central and Western Europe's most visible and vilified minority, became over the next century a topic for scientific debate. So widespread was the idea that the Jews differed from Europeans in some essential way—an explanation for the longevity of their civilization—that the notion of Jewish pathological uniqueness crept into the discourse of both friends of the Jews, such as Henri Baptiste Grégoire in France, and their bitterest enemies, such as Karl Wilhelm Friedrich Grattenauer in Germany.

Modern antisemitism, an ideology characterized by its denigration of the Jews on the basis of race, was born in the late eighteenth century, although clear manifestations of the belief that the Jews displayed a unique pathology are to be found in the Middle Ages. Irrespective of when racial antisemitism can be said to have begun, it is clear that the attempt to isolate Jews by labeling them as inherently different coincided with their political emancipation in the eighteenth century and their subsequent entrance into European society over the course of the nineteenth century.

By the late nineteenth century, race science, through its authority and prestige, bolstered traditional antisemitism by declaring that Jews possessed certain physiological traits that distinguished them from gentiles. The literature that dealt with such themes presented an array of answers to the Jewish racial question, and among its authors were Jewish race scientists. Their attempts to solve the supposed Jewish problem from a Jewish perspective should be regarded as a unique post-emancipatory venture in Jewish self-definition and self-assertion. The enterprise of the Jewish scientists was impelled by the external force of antisemitism and the internal need to reassert a Jewish ethnic pride that had been battered by the winds of assimilation.

At first glance, the thought of Jews conducting large-scale statistical experiments to determine Jewish skull shape and the prevalence of blue-eyed blond Jews seems a bitter irony. Our post-Auschwitz sensibilities do not allow that Jews themselves could have been part of such an enterprise. But the fact is that race science was central to the fin-de-siècle Western intellectual tradition. Modern European race thinking was not merely part of the ideology of extremist politicians and professional antisemites, or just a convenient tool for nationalist, antisemitic demagogues. The notion that races existed and were fundamentally different from one another, in a physical or psychological sense, was an integral element of modern European culture. It was a belief propounded with equal conviction by the aristocrat and the day laborer, the factory worker and the industrialist, by the illiterate masses and Europe's most gifted intellectuals. It was also an idea entertained by Jews. Educated Jews, as "good" acculturated Europeans, could not but believe in the concept of race. This was especially so for Jewish scientists, most of whom were trained as doctors in Europe's finest medical schools, where they had been exposed to curricula that were inherently racist.

Nevertheless, while Jewish and non-Jewish scientists clearly shared the same intellectual discourse about the various races of mankind, a fundamental difference of emphasis drove a wedge between their two programs. The scientists asked to what extent race, as a fixed set of physical and mental characteristics, limited a people's plasticity and adaptability, their potential for change? And what was the role of the environment, they asked, in creating the Jewish condition or, to put a finer point on it, the

Jewish racial problem? The evidence presented here clearly suggests that Jewish scientists laid the greater stress on the power of the environment, although their neo-Lamarackianism allowed them also to reserve a place for race, to entertain, for example, the idea that the environmentally determined aptitude of Jews for intellectual pursuits had over the tens of centuries become a racially transmissible characteristic.

To be sure, there were many non-Jewish race scientists who also shared the conviction that nurture was more important than nature. But even among them there was ambivalency and doubt when the Jews were involved. Well-intentioned humanists who believed in the singular power of the environment to effect physiological change, such as Johann Friedrich Blumenbach or Rudolf Virchow, were unable to shake off ancient suppositions about the essential nature of Jewish separateness. European, especially German, medical practice and research mirrored historic misgivings about Jews, according them a prominence in medical and anthropological discourse that was wholly out of proportion to their demographic and societal marginality. The German obsession with the Jewish racial question and the inability of the English to even acknowledge that one existed make clear the link between the comparative nature of Jewish integration and scientific inquiry in these two lands.

For the Jewish scientists examined here, the adoption of racial science to create a paradigm of Jewish self-definition was extremely complex. On the one hand, they largely rejected the anthropological interpretation of the Jewish problem, namely, that race was the driving force behind such things as Jewish existence, behavior, and superiority or inferiority. According to them, the physiological and psychological characteristics of Jewishness were primarily determined by history. And they observed two distinct historic forces pressing on the Jews—one external, the other internal. A poor physique or neurasthenic temperament was said to have been brought on by the debilitating effects of ghetto existence or antisemitism. These characteristics showed the impact of gentile history on the Jews. On the other hand, Jewish history, that is, the structure of Jewish life as created for and by Jews, made for their survival. All the positive aspects of Jewish life were the product of internal development. For the Jewish scientists, the category of race alone was insufficient to explain the longevity and historic greatness of Jewish civilization. To have accepted race as the sole determining factor would have of necessity obligated Jewish race scientists to accept entirely the terms of the general scientific debate. Increasingly, even secular scientists emphasized that to be Jewish was to belong to a special religious community. Moreover, as a corollary to the attempts of the Jewish scientists to create a new mode of Jewish self-definition, they valorized the traditionally observant and pious life of East European Jewry, seeing it not as an atavistic mode of existence but rather as one that could be exemplary for future generations of Jews, even those in the West. Ironically, it was secular Zionists in Germany, most of whom were long estranged from Jewish ritual, who repeatedly touted the prophylactic nature of traditional Judaism. Having identified the religious institutions of Jewish life as the principle factor in determining Jewish survival, Jewish race scientists employed the language of modern race science to refute many of its fundamental propositions concerning the Jews. They concluded that it was the fructifying qualities of Jewish culture, ethical and ceremonial, which made for the flowering of Jewish life through the ages. History, not biology, left the most significant impress on all peoples.

Nonetheless, Jewish scientists were hesitant to ascribe all positive Jewish features to the forces of history and culture. As men who believed in the concept of race, they often resorted to it as an explanation for some Jewish condition or other. For example, Joseph Jacobs, whose belief in the purity of the Jewish people was unshakable, was unable to assign the cause of the greater fertility of Jewish marriages to anything but racial factors. He offered a similar explanation for the seemingly younger age at which Jewish females began to menstruate. If race was a meaningless category, or a mere rhetorical device, why did Samuel Weissenberg devote an entire career to the gathering of anthropometric data about various little-known Jewish communities? When he visited these communities, he was primarily interested in data concerning their physical composition rather than their cultural output. For him, they were noteworthy mainly from a racial point of view, for they were integral links in the chain that would lead him to the *Urjude*. Ignaz Zollschan, although he eventually rejected the concept of race altogether, believed passionately in the idea early in his career. For him, there was something about the consistently stellar contributions that Jews had made to history and civilization, from so many places and over such a long period of time, that could best be attributed to their "racial genius." While the seeming absence of certain diseases, such as alcoholism, was attributed by Jewish physicians initially to social causes, many also noted that over the centuries the immunities had become hereditary. Thus, Jewish race scientists, like their non-Jewish counterparts, seemed to have been unsure as to where the influence of race began and where it ended, and at what point history became the decisive factor.

Although race was a meaningful category for the Jewish anthropologists, all of them warned sternly against the use of comparative physical anthropology to create a strident, racial chauvinism. Instead, they were impelled by the need to provide an authoritative picture of the sociology and anthropology of the Jews in order to defend against antisemitism, curb assimilation by inculcating a sense of ethnic pride among Jews estranged from their heritage, and win the respect of gentiles who touted the superiority of their own Christian culture and race.

In so doing, Jewish and non-Jewish anthropologists writing on Jews shared an important methodological assumption: that one could compose a unitary picture of "the Jew." Most treated the Jews as a group, en masse, undifferentiated by class, geography, and politics. Almost none of the scientists attempted to classify subgroups within the Jewish people, and when this was done, the effort only extended as far as distinguishing Sephardic from Ashkenazic Jews or East European from West European Jews. Thus the language of the Jewish race scientists also allowed for the stereotyping of Jewish existence. It was as though all East European Jews were observant, or all West European Jews were middle-class and estranged from their Jewish identities. Reading the scientists' texts rarely gives us a sense that social differences existed within a particular Jewish community. Because of their adoption of the gentile discourse and methodology on race, Jewish race scientists were unable to escape race science's natural inclination to create racial and ethnic typologies.

During the interwar period, the intellectual edifice of race science had begun to show cracks. Scientists in Britain and the United States had started to retreat from Nordic race theory, and the Nazi seizure of power only hastened the process. Not only did the international political situation spell the beginning of the end of scientific racism, but developments in science itself also made a significant contribution. The growth of a mathematically based population genetics served to highlight the nonempirical, epistemologically doubtful basis of traditional race science and eugenics. In Britain, prominent scientists such as Lancelot Hogben, J. B. S. Haldane, and Julian Huxley criticized eugenicists for their failure to acknowledge the crucial role played by the environment in genetics.

By contrast, in the United States, the rejection of scientific racism proceeded more slowly. The obvious and very different natures of American and British racism were reflected in the more strident bigotry to be found in the discourse on race in the United States. Public declarations of an antiracist position from within the scientific community did not become pronounced until after World War II. In the 1930s, the most vocal opponent of scientific racism was the Columbia University anthropologist Franz Boas. Himself a German Jew, Boas sought to organize a research program designed to refute scientific racism. Given the state of race relations in general, and the extent of antisemitism in American society in particular, Boas was concerned that his Jewishness would make his science appear partisan and thus compromised. In 1933, to challenge scientific racism, he enlisted the assistance of non-Jewish scientists such as Robert MacIver, Leslie Dunn, and A. T. Poffenberger. In addition to this project, the Boasian school of anthropologists, which included Alfred Kroeber, Robert Lowie, Melville Herskovits, Ruth Benedict, Margaret Mead, and the psychologist Otto Klineberg, played a central role in challenging scientific racism.

The older physical anthropology began to make room for what in Britain was called social anthropology and in the United States cultural anthropology. With the professionalization of the social sciences after World War I, sociologists, psychologists, ethnologists, and even many physical anthropologists who had become disaffected with the methodological inconsistencies and unempirical findings of race science focused on the dominance of environmental as opposed to biological factors to explain human society.

Jewish race science, with its emphasis on culture, had anticipated many of these developments. But at the same time its reliance on anthropometrics and descriptive racial classification meant that it too would have to give way. By contrast, just as these changes were taking place on both sides of the Atlantic, German science cooperated in the establishment of the racial state, which then proceeded to murder millions in the name of German racial superiority. The rise of Hitler also meant that Jewish race science had become a useless weapon against antisemitism and assimilation, and the battleground for Jewish self-defense and national rebirth could no longer be European, scientific journals. Nevertheless, in its heyday at the turn of the century, Jewish race science, although unable to dispel the malicious opinions of the antisemites, was, like all apologia written by Jews, able to offer Jewish readers comfort, dignity, and hope.

❋ NOTES ❋

Works frequently cited in the notes have been identified by the following abbreviations:

AA	*Archiv für Anthropologie*
ARGB	*Archiv für Rassen- und Gesellschaftsbiologie*
AZJ	*Allgemeine Zeitung des Judentums*
LBIYB	*Leo Baeck Institute Yearbook*
MAGW	*Mitteilungen der Anthropologischen Gesellschaft in Wien*
MGJV	*Mitteilungen der Gesellschaft für jüdische Volkskunde*
MGWJ	*Monatsschriftfür Geschichte und Wissenschaft des Judentums*
PAR	*Politisch-Anthropologische Revue*
TJHSE	*Transactions of the Jewish Historical Society of England*
ZDSJ	*Zeitschrift für Demographie und Statistik der Juden*
ZE	*Zeitschrift für Ethnologie*

One: Introduction

1. Notable exceptions are the work of Rossiter, *Women Scientists in America*, and Morantz-Sanchez, *Sympathy and Science.*

2. Nancy Leys Stepan and Sander L. Gilman, "Appropriating the Idioms of Science: The Rejection of Scientific Racism," in Dominick LaCapra, ed., *The Bounds of Race* (Ithaca: Cornell University Press, 1991), 72–103.

3. Increasing numbers of minority students at the universities have added a

sense of urgency and articulation to these debates. The current discourse in favor of multiculturalism takes as its starting point for discussions on curricula the right to challenge "how the relationship between culture and social identity is constituted "through hierarchical knowledge and power relations within the curriculum." Giroux, "Post-Colonial Ruptures," 11.

4. For example, see Clifford and Marcus, eds., *Writing Culture*.

5. Clifford, *Predicament of Culture*, 22.

6. For example, the pro-emancipationist Christian Wilhelm Dohm's *Über die bürgerliche Verbesserung der Juden* (1781) reflects the eighteenth-century conception of Judaism as an ethnoreligious designation. On the Maskilic meaning of the term *Jewish nation*, see Elbogen, "Die Bezeichnung 'Jüdische Nation.'" The historian Isaac Markus Jost's use of the term *Israelites* in his nine-volume *Geschichte der Israeliten* (1820–1828) reflected his personal antipathy to rabbinic Judaism and the fact that he considered it an historical abberation. For Jost, emphasis on the Israelites harked back to a more noble and pristine time in the history of Judaism, one he believed modern Jewry should emulate. See Schorsch, "Wolfenbüttel to Wissenschaft," 118. This was the official name of the largest Jewish organization in modern Germany, the Centralverein deutscher Staatsbürger jüdischen Glaubens (C.V.). The issue of changing nomenclature is an important one in modern Jewish history, going to the heart of oscillating modes of self-perception. On the ideological implications of the different names Jews used at varying times to refer to themselves, see Meyer, *Response to Modernity*, 30.

7. See especially Stepan, *Idea of Race in Science*; and Mosse, *Toward the Final Solution*.

8. Trachtenberg, *Devil and the Jews*, 50 and 149; Poliakov, *History of Anti-Semitism*, 1:143, nn13,14; Yerushalmi, *Spanish Court to Italian Ghetto*, 126–133; and Gilman, *Jewish Self-Hatred*, 74–75.

9. Yerushalmi, *Assimilation and Racial Anti-Semitism*.

10. Grégoire, *Reformation of the Jews*, 35–36. Hereafter, references to Grégoire are cited in the text.

11. Grégoire had also said this. Of Jewish men he wrote, "They have almost all red beards, which is the usual mark of an effeminate temperament" (56).

12. See, for example, Beadles, "Insane Jew,"; and the response by Benedikt, "Open Letter to Beadles."

13. The distinguished German psychiatrist Emil Kraepelin declared that Jews suffered from an extraordinarily high incidence of mental illness, brought on by hereditary degeneration. According to Kraepelin, this was most likely caused by the Jewish preference for consanguineous marriage. See Kraepelin, *Psychiatrie*, 1: 106.

14. Katz, *Prejudice to Destruction*, 292–300.

15. Mosse, *Toward the Final Solution*, 123.

Two: German Race Science: The Jew as Essential Other

1. For example, see Curtin, *Image of Africa*.
2. A telling example is the Huron hero of Voltaire's *L'Ingénu*. At first he is praised for his beauty and then later for speaking excellent French. He is, of course, not commended for speaking his native tongue. By the end of the novel, the Indian is lauded not because he is a noble savage but because he had moved to Paris, lived and dressed like a Frenchmen, assumed a different name, and achieved rank as an officer in the king's army. The concept of the noble savage must have been, in Voltaire's mind, a pure abstraction and fantasy, because *l'ingénu* turned out to be a Frenchman from the very beginning. Even had he been a genuine Huron, Voltaire had merely honored him because of his European qualities and never for his own ethnicity. See Buchanan, "Savages, Noble and Otherwise."
3. Querner, "Zur Geschichte der Anthropologie."
4. Lovejoy, *Great Chain of Being*, esp. 227–241; and Stepan, *Idea of Race*, 1–19.
5. Blumenbach also noted that "for physiological reasons," such as the Georgian's beauty and whiteness, this mountainous region was most likely the home of original man. Blumenbach assigned to the Georgian a centrist position between the two "ultimate [human] extremes, on the one side the Mongolian, on the other the Ethiopian." He hypothesized that "we may fairly assume [white] to have been the primitive colour of mankind, since . . . it is very easy for that to degenerate into brown, but very much more difficult for dark to become white." Blumenbach, *On the Natural Varieties of Mankind*, 269.
6. Quoted in Querner, "Anthropologie," 289. For more on Meiners and other Enlightenment anthropologists who distinguished between beautiful and ugly races, see Poliakov, *Aryan Myth*, 155–182.
7. Mosse, *Toward the Final Solution*, 22.
8. Klemm, *Allgemeine Cultur-Geschichte der Menschheit*, 1:196–197.
9. Weerth, *Entwicklung der Menschen-Rassen*, 87–88.
10. Preston, "Science, Society, and the German Jews," 71.
11. On the development of this Jewish subculture, see Volkov, *Jüdisches Leben und Antisemitismus*, 131–145; and Sorkin, *Transformation of German Jewry*.
12. Ploetz, *Tüchtigkeit unserer Rasse*, 130–142.
13. Mendes-Flohr and Reinharz, *The Jew in the Modern World*, 103–105. On the attitude of the German left to Jews and Judaism, see Wistrich, *Socialism and the Jews*.
14. Ploetz, *Tüchtigkeit unserer Rasse*, 142.
15. Poliakov, *Aryan Myth*, 295.
16. See, for example, the following articles in *Politisch-Anthropologische Monatsschrift* 19, no. 4 (1920–1921): "Die groe Völkertäuschung durch die Juden," 160–164; "Eine Parallele zum Judentum," in which the author writes, "Das

Judentum spielt völkerbiologisch die Rolle gewisser Fäulnisbakterien; wer das nicht einsieht, dem ist nicht zu helfen," 182–184; and no. 8, "Geist und Judentum," 418–423.

17. See Proctor, *Racial Hygiene*, 15; and Weiss, *Race Hygiene and National Efficiency*, 2.

18. William Nussbaum file, Leo Baeck Institute, New York, LBI/AR-725.

19. On Fischer: Proctor, "From *Anthropologie* to *Rassenkunde*," 146; Lenz: Proctor, *Racial Hygiene*, 53–55.

20. Blumenbach, *Natural Varieties*, 234.

21. Wachter, "Bemerkungen über den Kopf des Juden," 64–65.

22. Rudolphi, *Beyträge zur Anthropologie*, 153.

23. Ibid., 159. Such views were not, of course, confined to Germany. In 1829, the Anglo-French founder of the Société Ethnologique de Paris, William Frederick Edwards, sent the liberal historian Amédée Thierry a letter in response to the latter's publication in 1828 of his *Histoire des Gaulois*. In the letter, Edwards noted that "peoples which have settled in different climates are able to preserve their distinctive types for centuries." Quoted in Poliakov, *Aryan Myth,* 226. It seemed to Edwards that the Jews, because of their dispersal and omnipresence, were the perfect subjects to study in order to prove conclusively that the Buffonian notion of human diversity, resulting from the effects of "climate, nutrition and the prevailing way of life," was incorrect. A study of the Jews would also prove that the corollary of Buffon's theory, the Lamarckian notion of the inheritance of acquired characteristics, could not take place. Edwards concluded, after an examination of pictorial evidence, which included Leonardo da Vinci's *Last Supper* and the tomb of an Egyptian pharaoh on display in London, that the characteristics of the Jews depicted on the tomb show "a 'striking' resemblance to those of the modern Jewish population of that city." Quoted in Blanckaert, "On the Origins of French Ethnology," 35. For Edwards, the Jewish type was indelible.

24. Andree, *Zur Volkskunde der Juden*, 24.

25. Ibid., 37.

26. Stepan, *Idea of Race in Science*, 97.

27. Vogt, *Lectures on Man*, 433.

28. Maurer, "Mitteilungen aus Bosnien."

29. Weisbach, "Körpermessungen verschiedener Menschenrassen," 214.

30. Blechmann, *Beitrag zur Anthropologie der Juden*, 59.

31. See Stieda, "Beitrag zur Anthropologie der Juden."

32. See Kollmann, "Schädel und Skeletreste," 650–651.

33. For his theory of racial unchangeability, see Kollmann, "Rassenanatomie der Hand."

34. Ikow, "Neue Beiträge zur Anthropologie der Juden," 15.

35. Lagneau, "Sur la race juive et sa pathologie."

36. Virchow, "Gesamtbericht . . . über die Farbe."

37. Mosse, *Toward the Final Solution*, 90.

38. Claiming that Virchow was Jewish, the antisemites attacked him, taking great umbrage at his findings that the Germans were not a pure race and that Jews possessed Aryan racial characteristics. See Kümmel, "Virchow und der Antisemitismus."

39. Alsberg, *Rassenmischung im Judentum*, 7.

40. Luschan's theory is discussed more fully in chapters 4 and 5 below.

41. One of the most important treatises elaborating this view was presented in 1895 by Georg Buschan, a private practitioner from Stettin and the editor of the *Zentralblattes für Anthropologie*. Buschan, who denied Jewish racial heterogeneity, argued that pathological peculiarities of Aryans and Semites were essentially different. See Buschan, "Einfluß der Rasse."

42. See Fishberg, *Race and Environment*, 270–355.

43. Kraepelin, *Psychiatrie*, 1:106. For a fuller explanation of Kraepelin's degeneration theory, see his "Zur Entartungsfrage."

44. Krafft-Ebing, *Text-Book of Insanity*, 143. In keeping with the ambivalent attitude of the German medical profession toward Jews, Krafft-Ebing, despite his views, could treat individual Jews rather differently. In 1897, together with Hermann Nothnagel, head of the Division of Internal Medicine at the University of Vienna, he proposed Sigmund Freud for the position of *Ausserordentlicher Professor*. Individual Jews were always more acceptable than Jews en masse.

45. On Jewish psychiatrists' evaluation of mental illness among the Jews, see Efron, "'Kaftanjude' and the 'Kaffeehausjude.'"

46. Mosse, *Crisis of German Ideology*, 149–233.

47. Tal, *Christians and Jews in Germany*.

48. This figure refers only to German-born Jews. When foreign Jews are included, the picture changes even more dramatically. In 1911, for example, there were more foreign Jews studying medicine at the University of Königsberg than native-born Jews. Furthermore, in 1891 foreign Jews made up 7 percent of all Jewish students; by 1911 this figure had jumped to 56 percent. Preston, "Science, Society, and the German Jews," 105–106. I rely on Preston's figures for this discussion.

49. Ibid., 108–112; and Breslauer, *Zurücksetzung der Juden*.

50. Quoted in Klein, *Jewish Origins of the Psychoanalytic Movement*, 51.

51. Ehrke, "Antisemitismus in der Medizin."

52. Lewy, "Antisemitismus und Medizin," 6.

53. Ibid., 8.

54. Ibid., 7.

Three: British Race Science: The Jew as Non-Other

1. Panitz, *Alien in Their Midst*, 162. While Panitz's view is essentially borne out by other scholars, a far more nuanced account of the problem is Fisch, *Dual Image*. See also Stone, "From Fagin to Riah: Jews and the Victorian Novel."

2. On the origins of British anthropology and ethnology and their changing intellectual, political, and institutional concerns, see Stocking, "What's in a Name? The Origins of the Royal Anthropological Institute."

3. Endelman, *Jews of Georgian England*, 119–192.

4. Stocking, *Victorian Anthropology*, 9–45.

5. For a discussion of the commercial incentive to readmit the Jews, see Roth, *History of the Jews in England*, 149–172.

6. On the myth of the Lost Tribes and their relation to the readmission of the Jews to England, see Katz, *Readmission of the Jews*, 127–157; Hyamson, "Lost Tribes.".

7. Williams, ed., *Adair's History of the American Indians*.

8. For a brilliant analysis of this subject, see Leon Poliakov, *Aryan Myth*, 1–128.

9. Roth has also pointed out that cordial relations were forged because of a number of concrete historical factors: the close ties, from resettlement on, between England's equivalent of Germany's court Jews and the court and government circles; occupational restructuring, which saw Jews enter the theater and musical worlds; the success enjoyed by world-class Anglo-Jewish pugilists in the late eighteenth century, which proved to the British public that the Jew was not in need of physical improvement, as was so often declared on the Continent; the rise of the Evangelical movement, which bred a certain tolerance; and a slow and gradual acceptance and incorporation of Jews that never outstripped public opinion. Roth, *History of the Jews*, 239–267.

10. Toland, *Reasons for Naturalizing the Jews*, 26–27.

11. Ibid., p. 30.

12. Quoted in Poliakov, *Aryan Myth*, 44.

13. A classic example of this kind of thinking is Poole, *Anglo-Israel*.

14. Buchanan, *Christian Researches in Asia*, 169–176.

15. In France, the observation that the complexions of the Jews approximates those of their neighbors had already been made by François-Maximilien Misson in 1691 and Georges Louis Leclerc, comte de Buffon, in his *Natural History* in 1749—long before Buchanan called Jewish skin color to the attention of the British.

16. On this issue, see Gould, *Mismeasure of Man*.

17. Nott and Gliddon, *Types of Mankind*, 120.

18. Stocking, "What's in a Name?," 373.

19. Stepan notes that the initial goal of Prichard's anthropology was to "reassert the traditional Chrisitan view of the unity of the different races of mankind." *Idea of Race*, xiii.

20. See Stocking's introductory essay to Prichard, *Physical History of Man*, ix–cx.

21. Prichard, *Physical History of Man*, 176. Hereafter references to this work appear in the text.

22. Smith, *Causes of the Variety of Complexion*, 42.

23. Gilman, *Freud, Race, and Gender*, 49–69.
24. On Lawrence, see Stepan, *Idea of Race*, 1–39.
25. Lawrence, *Lectures on Physiology*, 437.
26. Ibid., 468–469.
27. On the idea of European history as characterized by the conquering of autochthonous populations (represented by the lower classes) by invading warriors (represented by a pan-European aristocracy), see Hannah Arendt, *Origins of Totalitarianism*, 161–175.
28. Stocking, "What's in a Name?," p. 373.
29. Latham, *Natural History*, 471.
30. Ibid., 514–515.
31. On Gobineau, see Biddiss, *Father of Racist Ideology*. Too many historians have uncritically accepted Gobineau as the "father of racist ideology," exaggerating his importance as a thinker about racial difference as well as the extent of his influence on European ideas of race in general. He has even been seen by some as an ideological precursor of the Nazis. George L. Mosse has correctly located Gobineau in the history of European racial thought in his *Toward the Final Solution*, 51–62. One of the finest analyses of Gobineau's *Essai* and its contemporary impact is in Jacques Barzun, *Race: A Study in Modern Superstition*, 50–77.
32. On Knox and the general development of anti-environmentalist anthropology after 1850, see Stepan, *Idea of Race*, 41–46; Stocking, *Victorian Anthropology*, 62–69; Rae, *Knox, the Anatomist*; and Curtin, *Image of Africa*. Curtin suggests that the first important proponent of a thoroughly racial view of history in Great Britain was Dr. Robert Knox, whom he calls "the real founder of British racism and one of the key figures in the general Western movement toward a dogmatic pseudo-scientific racism" (377). A contrary view is in Henry Lonsdale, *Life and Writings of Robert Knox* (1870). Lonsdale, a student of Knox, lamented the fact that his teacher was primarily remembered as an anatomist and not for his works on race. He predicted that "posterity will probably figure him as the chief anthropologist of his epoch, and a pioneer of a philosophy that sought to recognize the true nature of Man, his instincts, his passions, his psychological leanings, and social influences" (330).
33. Ibid., p. 46.
34. Douglas, "Burke and Hare."
35. Greene, *Science, Ideology, and World View*, 60–94.
36. Knox, *Races of Men*, 10. References to this work appear in the text.
37. Disraeli, *Coningsby*, 271.
38. Ibid., 274.
39. Recent reassessments of Disraeli's attitude to his Jewish identity argue that he was more hostile toward it than is commonly assumed. The best treatment is Todd M. Endelman, "Disraeli's Jewishness Reconsidered." Taking the lead from Hannah Arendt, Endelman is cognizant of the real importance of race for Disraeli. He correctly sees Disraeli's affiliation to his Jewishness as strictly biological and

therefore devoid of anything philosophically or spiritually Jewish. Also see Arendt, *Origins*, 72–75. A superb discussion of the psychological implications of Disraeli's Jewish identity is Isaiah Berlin, "Disraeli, Marx and the Search for Identity.".

40. For a discussion of the accusation of artistic plagiarism by Jews, see Gilman, *Jewish Self-Hatred*, 209–211.

41. For a detailed but flawed analysis of *Das Judenthum in der Musik*, see Jacob Katz, *Darker Side of Genius*. Katz incorrectly denies the racial base of Wagner's antisemitism.

42. Wagner, *Prose Works*, 3:84.

43. The need to create identifying marks relates to the role of stereotyping or creating an idealized world where groups who promote fear or anxiety, in this case assimilated Jews, become the objects of a stereotyped abstraction with a particular set of characteristics by which that group can be easily identified. According to the stereotype, Jews could often be identified by their blackness, sexual promiscuity, and big noses. For a general discussion of the role of stereotyping, see Gilman, *Difference and Pathology*, 15–35.

44. Marx, *Marx-Engels Werke*, 30:257–259. On Wagner, see Stern, *Gold and Iron*, 242–243, 498. For a discussion of the Jew as black, see Gilman, *Jewish Self-Hatred*, 6–12.

45. For example, the *völkisch* ideologist Julius Langbehn summed up the contempt in which assimilated German Jews were held: "Rembrandt's Jews were real Jews, who wanted to be nothing other than Jews, and they also had character. This is the exact opposite of today's Jews; they want to be Germans, Englishmen, Frenchmen etc., and through this have become characterless." Langbehn, *Rembrandt als Erzieher*, 43.

46. Graetz, *Structure of Jewish History*, 239; quoted in Graetz's paper "Historical Parallels in Jewish History," in *Papers Read at the Anglo-Jewish Historical Exhibition* (1888).

47. See *Dictionary of Scientific Biography*, ed. Gillispie (New York, 1970), 1: 562–563.

48. Beddoe, "On the Physical Characteristics of the Jews," 236.

49. On these events and their profound effects on contemporary self-perceptions in Britain, see Bolt, *Victorian Attitudes to Race*; Searle, *Quest for National Efficiency*; and idem, *Eugenics and Politics in Britain*.

50. See Cowen and Cowen, *Victorian Jews through British Eyes*. As is well known, popular resentment in England towards the Irish at this time far outstripped any antipathy towards Jews.

Four: Joseph Jacobs and the Birth of Jewish Race Science

1. For Aufklärung views of the Jews in Germany, see Low, *Jews in the Eyes of Germans*; Rotenstreich, *Studies in Anti-Judaism in Modern Thought*; Katz, *Out of the Ghetto*; and the excellent work concerning anti-Jewish attitudes over eman-

cipation in Germany of Rainer Erb and Werner Bergmann, *Die Nachtseite der Judenemanzipation*. For France, see Hertzberg, *The French Enlightenment and the Jews*, and the still valuable article of Isaac Barzilay, "The Jew in the Literature of the Enlightenment."

2. As yet, there is no complete bibliography of Jacobs's works. His published books, scholarly articles, journalistic contributions, book reviews, and encyclopedia entries in such areas as anthropology, folklore, literary criticism, and Jewish history, literature, and philosophy amount to well over eight hundred items.

3. Rutland, *Edge of the Diaspora*, 50.

4. Benjamin, "Joseph Jacobs," 75.

5. Wolf later described this as "a very large part of the sunshine of my life." See Lucien Wolf's address to the memorial meeting for Jacobs in *TJHSE* (1918): 148.

6. *Alumni Cantabrigienses, Part 2, 1752 to 1900*, vol. 3 (Cambridge, 1947), 543.

7. Jacobs, *Jewish Ideals*, xii–xiii. Eliot's novel made a noticeable impact on contemporary Jews. See Kaufmann, "George Eliot und das Judentum."

8. Jacobs, *Jewish Ideals*, xii. Thomas Henry Huxley was lecturer in natural history at the Royal School of Mines and is remembered as the main advocate of Darwin's theory of evolution. William Kingdon Clifford was professor of applied mathematics at University College, London. He played a central role in introducing the ideas of non-Euclidean geometers to English mathematicians. He was also a staunch supporter of Darwinian evolution.

9. Jacobs was, of course, not blind to British antisemitism. Not only did he sarcastically declare that "we have Englishry defended from Jewry by a gentleman of the un-English name of Hillaire Belloc," he also had his friend, the Anglo-Jewish author Israel Zangwill, arrange a meeting, or rather a confrontation, between himself and the antisemitic author G. K. Chesterton. Zangwill recalled Jacobs's "unjaded intellectual curiosity to study the new Anti-Semitic wing of English letters, which had grown up so oddly in his [Jacobs's] absence, he got me to bring him together with Mr. G. K. Chesterton. The meeting took place at the 'Cheshire Cheese,' where if the duel of Aryan and Semite came off without casualties, it was perhaps because I had prudently made provision in the spirit of one of the Tosaphoth of Rabbi Elchanan translated by Jacobs in *The Jews of Angevin England*. 'It is surprising,' comments the worthy Rabbi, 'that in the land of the isle the Jews are lenient in the matter of drinking strong drinks of the Gentiles and along with them. . . . But perhaps as there would be great ill-feeling if they were to refrain from this, one must not be severe on them.'" See Zangwill's address to the memorial meeting for Jacobs in *TJHSE*, (1918): 131–132. On British antisemitism, see Holmes, *Anti-Semitism in British Society*.

10. The issue of mounting a dignified defense of the Jewish people was of great importance to Jacobs. In his attack on Chamberlain's *Foundations of the Nineteenth Century* (1899), Jacobs declared that the author's "arrogance and in-

solence of the exclusive claims that it advances on behalf of the 'Teutonic' genius
... exceeds all the bounds of good taste and even good breeding." Dignity and
biting wit would become hallmarks of Jacobs's own apologia. See his essay "The
Higher Anti-Semitism" in his collection *Jewish Contributions to Civilization*, 51.

11. See Jacobs, *Jewish Contributions*, 44–45 and 49–51.

12. Among Jacobs's several hundred publications, he made seminal contributions to the study of pre-expulsion English Jewry, Spanish Jewish history, and biblical archeology. He also played a leading role, as a founder of the Jewish Historical Society of England and secretary of the Society of Hebrew Literature, in the institutionalization of Jewish scholarship in Great Britain. Together with Lucien Wolf, he organized the great Anglo-Jewish Historical Exhibition of 1887, compiling its catalogue and a large bibliographical source book for the study of Anglo-Jewish history, *Bibliotheca Anglo-Judaica*.

13. *Jewish Chronicle*, February 5, 1982, p. 20. In practical terms his contribution to the Russian Jews was considerable. Aside from a remarkable little book entitled *The Persecution of the Jews of Russia* (1890), in which he gave a vivid historical account of Russian Jewry and a lucid summary of the various anti-Jewish laws in Russia, he co-authored with Hermann Landau the *Yiddish-English Manual* (London, 1893) which went through six editions by 1906 and was a standard text for the thousands of Russian Jewish immigrants struggling with the English language.

14. Jacobs, *Jewish Statistics*, preface.

15. See his lecture delivered before the Ethical Society on May 12, 1889, entitled "Jewish Ideals," in his *Jewish Ideals*, 1–23.

16. Jacobs made the speech to the Royal Academy on the occassion of his election to that body as a member. Jacobs's work in the Spanish archives led to his publishing *An Inquiry into the Sources of the History of the Jews in Spain* (1894). His speech is reprinted in *Jewish Ideals*, 234–242.

17. Jacobs, in *A Jewish Scholar's Career: The Maccabaeans*, 8.

18. Jacobs's most important anthropological writings on the Jews, done between 1882 and 1889, were gathered together and published in a single volume entitled *Jewish Statistics: Social, Vital and Anthropometric* (1891). It is this edition of his works that has been used in the preparation of this chapter.

19. Jacobs's works on the subject, especially on the poor among the Jewish population of England, were considered path-breaking and are still of use to students of Anglo-Jewish social history. They significantly predate Charles Booth, ed., *The Life and Labour of the People of London* (1889–91), with its important essays on London Jewry by Beatrice Potter and on immigration by Sir Hubert Llewellyn Smith. Similarly, Jacobs's research appeared before the two important government studies that concerned themselves with the Jewish immigrants, *Report of Select Committee of House of Commons on Immigration and Emigration* (1888) and *Report of House of Lords Select Committee on Sweating System* (1888–89). The eminent Anglo-Jewish historian Vivian D. Lipman has described

Jacob's essays as "brilliant." For a sociological study of English Jewry as well as the historiography on that subject, see Lipman, *Social History of the Jews in England.*

20. On the anti-immigrationists, see Feldman, "The Importance of Being English."

21. Polygamy had long been taboo for European Jewry. The prohibition against bigamy for the Jews of the Rhineland had, since the twelfth century, been ascribed to Gershom ben Judah Me'or HaGolah (c. 960–1028). It seems, however, that the establishment of monogamy as a cornerstone of Jewish marital life was not the result of a single legislative act by Gershom, because polygamous marriages still occurred in his day. Rather, it was introduced slowly, by degrees. The practice had achieved universal acceptance by the time of the great Torah sage Rashi (1040–1105). See Falk, *Jewish Matrimonial Law,* esp. 1–34.

22. England was not alone in having a higher consanguineous marriage rate among its Jewish population. It seemed to be a European-wide phenomenon. For example, in Alsace-Lorraine, at the turn of the century, the proportion of consanguineous marriages per thousand marriages was 1.86 among Protestants, 9.97 among Catholics, and 23.02 among Jews. Fishberg, *The Jews,* 250. Similarly, a study of consanguineous marriage rates among Jews in Hungary between 1901 and 1906 showed that although Jews comprised only 4.4 percent of the total population in 1900, they had 27.55 percent of all consanguineous marriages. *ZDSJ* 8, no. 3 (1907): 46; and 9, no. 7 (1908): 110.

23. Jacobs, *Jewish Statistics,* 6.

24. Darwin's most important work on this topic as it pertains to plant life is *Effects of Cross and Self Fertilization.* He came to a similar conclusion about the evil effects of the inbreeding of animals. As far as man was concerned, Darwin hesitated, maintaining that if evil effects did follow from consanguineous marriages, they were probably "very small." See his *Variation of Animals and Plants under Domestication,* 2:102–104. A vast amount of work on this subject was done by medical practitioners, with little unanimity of conclusion. Some of the more important European texts declaring the harmfulness of consanguineous marriages include Boudin, *Traité de Géographie,* 2:131–142; idem, "Sur l'idiote et l'aliénation mentale"; Devay, *Du danger des mariages consanguins,* esp. 177–192; and Virchow, "Über Erblichkeit." Virchow found that in Berlin there was one deaf-mute in every 3,179 Roman Catholics, 2,173 Protestants, and 673 Jews. He concluded (354) that a causal link existed between these figures and the high incidence of consanguineous marriage among Jews.

25. Jacobs, *Jewish Statistics,* 7–8.

26. Among an earlier generation of scientists were Auguste Voisin, "Mariages entre consanguins" (1865); J. A. N. Perier, "Croisements ethniques," (1860–1863); and A. H. Huth, *The Marriage of Near Kin* (1875).

27. Jacobs, *Jewish Statistics,* 7.

28. The fullest expression of Jacobs's view of the Jewish religion, moral sys-

tem, and conception of God is his essay "The God of Israel," reprinted in *Jewish Ideals*. The novelist Israel Zangwill regarded this as Jacobs's "most brilliant and creative essay": "This essay, for the breadth of its sweep and the mingled boldness and sobriety of its conception, is unparalleled in literature as the work of a young man of twenty-four." See "Address by Mr. Israel Zangwill," *TJHSE* (1918): 133.

29. Jacobs, *Jewish Statistics*, 10. In fact, another researcher, the American Maurice Fishberg, noted: "When nearly every vestige of the traditional Jewish solidarity has vanished, as is the case in some communities in Western Europe, charity apparently remains, and the Jewish poor are kept from the door of the Gentile." Fishberg, *The Jews*, 360–361.

30. The literature on late nineteenth-century German antisemitism is enormous. On the most infamous of the financial scandals, that involving the converted financier and railroad builder Bethel Henry Strousberg, see Stern, *Gold and Iron*, esp. 352–369 and 494–531. On France, see Wilson, *Ideology and Experience*, who notes that the crash of the Union Générale in 1882 was widely attributed to the machinations of Jewish bankers (247–318). The best studies of political and economic antisemitism in Germany remain Pulzer, *Political Anti-Semitism in Germany and Austria*, and Massing, *Rehearsal for Destruction*. For an intellectual history of modern antisemitism, see Katz, *From Prejudice to Destruction*. The best account of the world Jewish conspiracy theory is in Cohn, *Warrant for Genocide*.

31. Jacobs drew attention to the dislocating effect of the rapid economic development of Germany and its role in the antisemitic movement in the final decades of the nineteenth century. See his "The Higher Anti-Semitism," in *Jewish Contributions*, 9–58.

32. Jacobs, *Jewish Statistics*, 10–11.

33. Ibid., 14.

34. Ibid., 18.

35. Jacobs, *Jewish Ideals*, 20–21. The idea of the Jews advancing the march to progress of the nations among whom they lived was shared by many contemporary intellectuals. For example, the French historian Ernest Renan wrote that "the Jew was designed to serve as leaven in the progress of every country, rather than to form a separate nation on the globe." Renan, *History of the People of Israel*, 5:9.

36. Jacobs, *Jewish Statistics*, 21.

37. For a summary of contemporary figures attesting to the greater vitality of Jewish over non-Jewish children, and that this was a result of Jewish religious practice, see "Report of a Lecture by B. W. Richardson."

38. The *Jewish Chronicle* of 1882 and 1883 provided extensive coverage of the case, and Jacobs's articles on Jewish statistics and the history of the ritual murder charge were carried in the same editions—often on the same pages.

39. *Jewish Chronicle*, June 29, 1883, pp. 6 and 8–9.

40. Jacobs, "Little St. Hugh of Lincoln."

41. Jacobs, *Jewish Statistics*, 22. Jacobs's statistics for the economic history of Anglo-Jewry are still exceptionally valuable, and though some of the figures have been emended, recent research tends to confirm most of his basic findings. See Pollins, *Economic History of the Jews in England*.

42. For example, see Erb, "Warum ist der Jude zum Ackerbürger nicht tauglich?"

43. Jacobs's article in defense of the varied occupational structure of European Jewry was preceded in 1880 by the work of a Protestant clergyman named Rost, who staunchly defended the Jews against charges that they were swindlers, workshy, and innately averse to manual labor. In this noteworthy little book, the author set out to test the truth and validity of the claims made about the working ways of the Jews by an historical analysis of what occupations Jews held over the centuries. See Rost, *Zur Berufsthätigkeit der Juden*, 5–6.

44. Jacobs, *Jewish Statistics*, 24.

45. Jacobs noted that "the 'Judaism of the Stock Exchange' turns out to be by no means an extraordinary percentage of that institution (only 5 per cent). The chief Jewish monopolies appear in cocoa-nuts, oranges and esparto grass, canes, slippers, sponges, and umbrellas, furs and ostrich feathers, meerschaum pipes and valentines—scarely the commodities which rule the world of prices." Ibid., 38. See also Pollins, *Economic History*, 171–182.

46. Jacobs, *Jewish Statistics*, 25–28. Nevertheless, commercial pursuits were taken up by Jews in numbers that exceeded their percentage in the total population, a fact which Jacobs and other contemporary Jewish researchers did not deny. See Jeiteles, *Die Kultusgemeinde der Israeliten in Wien*, 74–76.

47. Jacobs, *Jewish Statistics*, 24–25. This situation holds true to this very day. Among orthodox communities, occupational choice is to a great extent governed by the social and cultural demands of Judaism. For example, 54 percent of Lubavich men in Montreal work in religiously oriented occupations. Similar figures are available for New York. If not employed in jobs that are directly connected to the religion (e.g., the production of kosher food), many contemporary Orthodox Jews work in businesses owned by other members of their community. In this way, the work day can be structured so as to compliment the demands of ritual observance. See Sharot, "Hasidism in Modern Society," 518.

48. Jacobs, *Jewish Statistics*, 41.

49. Mosse, *Toward the Final Solution*, introduction.

50. Galton, *Hereditary Genius*, 23.

51. See Katz, *From Prejudice to Destruction*, 175–194.

52. Jacobs, *Jewish Statistics*, 41.

53. See Searle, *Quest for National Fitness*.

54. Jacobs, *Jewish Statistics*, 49–50; and "Marriage," in *Jewish Encyclopaedia*, 8:339.

55. Jacobs, *Jewish Statistics*, 50–53.

56. On comparative menstruation, see Raciborski, *Traité de la Mentruation*,

esp. 229–233. Raciborski found that menstruation made its first appearance among Jewish girls on average at 14 years, 3 months, 25 days, and in Slavic girls at 15 years, 3 months, 9 days. See also Weber, "Über die Menstrualverhältnisse der Frauen in St. Petersburg," and Weissenberg, "Die südrussischen Juden."

57. Jacobs, *Jewish Statistics*, 52.

58. The large-scale anthropometric surveys that Jacobs did in the coming years supported these early suspicions and were widely confirmed by other race scientists. For example, on the chest measurement of Jewish youths, see Blechmann, *Ein Beitrag zur Anthropologie der Juden*; Goldstein, "Des circonférences du thorax"; and Ammon, *Die natürliche Auslese beim Menschen*, 134.

59. Here Jacobs is alluding to Genesis 2:18: "And the Lord God said, It is not good that the man should be alone; I will make him an help meet for him."

60. *Jewish Chronicle*, March 20, 1885, p. 5. On the tremendous rate of Jewish population growth in the nineteenth century, see Lestschinsky, Brutzkus, and Segall, eds., *Bleter far yidishe demographie, statistik un ekonomik*, and Baron, "Ghetto and Emancipation," esp. 521–22.

61. In the revised English edition of Arthur Ruppin's *Die Juden der Gegenwart* (1904), which appeared in 1934, the author identified birth control, of which he was an advocate, as a major cause of the low Jewish birthrate in the West, and feared the consequences of its unrestrained use. "In certain countries birth-control among the Jews is so severe that their numbers decline, and there is the danger of gradual extinction; and though this applies to the non-Jews in the cities of Western and Central Europe, they at least, in contrast to the Jews, have considerable human reserves in their agricultural population." Ruppin, *Jews of the Modern World*, 76.

62. Some of the discrepancies between the number of births per Jewish marriage and those per gentile marriage were startling. For example, Austria (1851–75), 8.8 and 4.4; Austria (1861–70), 8.8 and 3.8; Baden (1857–63), 5.0 and 4.1; and France (1855–99), 3.9 and 3.0. Jacobs ascribed the unusually large difference between the number of Jewish and Christian children per family in Austria to the fact that a many Jewish marriages were not entered in civil registers. Although the difference would have been less, the Jewish figure would still have been higher than the gentile. The greater number of children to a Jewish marriage was not, however, universal. For example, the figures for Prague (1865–74) were 2.6 and 4.1; for Russia (1867), 3.8 and 5.0; and (significantly for the later researches of Theilhaber and Ruppin) Berlin (1881), 3.9 children per family for both Jews and Christians. See Jacobs's article "Births," in the *Jewish Encyclopaedia*, 3:224.

63. Jacobs, *Jewish Statistics*, 56.

64. Jacobs, "On the Racial Characteristics of Modern Jews," in *Jewish Statistics*, vi.

65. It will be recalled that earlier Jacobs had argued for the predominance of Jewish females over males. Contradictions such as this abound in the literature on race. On the predominance of male births, see Westermarck, *History of Human*

Marriage, 3:179–80; Nichols, "Numerical Proportion of the Sexes at Birth"; and Darwin, *Descent of Man,* 1:249.

66. See Hofacker, *Eigenschaften;* Platter, "Die Hofacker-Sadler'sche Hypothese"; and Nagel, "Der hohe Knabenüberschuss der Neugeborenen der Jüdinnen."

67. Jacobs, *Jewish Statistics,* 57–59.

68. Ibid., v.

69. *Jewish Chronicle,* February 27, 1885, p. 12.

70. Adolf Neubauer, "Race-Types of the Jews," 19.

71. Jacobs, *Jewish Statistics,* iii.

72. See the article "Expectation of Life" in the *Jewish Encyclopeaedia,* 5:306–308.

73. Jacobs, *Jewish Statistics,* vii. On the final point about connubial relations, Jacobs speculated fantastically in a footnote: "May this custom of separation (Lev. XV, 19) have any connection with Jewish proficiency in music, which in its origin seems to be also regulated sexual emotion?" Ibid., vii.

74. As an example of the loosening of tradition among Jews, Jacobs cited the rising rate of illegitimate births and suicides. Ibid., xi.

75. See, for example, Weissbach, "Körpermessungen verschiedener Menschenrassen"; Scheiber, "Untersuchungen über den mittleren Wuchs der Menschen in Ungarn"; Ranke, "Zur Statistik und Physiologie der Körpergrösse der bayerischen Militärpflichtigen"; Himmel, "Körpermessungen in der Bukowina," 83–84.

76. With an average height for European males (set by the Franco-Russian anthropologist Joseph Deniker) of 165 cm, Jacobs showed that Polish Jews averaged 161.2 cm, Galician Jews 162.3 cm, and Russian Jews 163.7 cm. But it was the measurements of upper-class English Jews by the British Association in 1883 that confirmed for Jacobs the continued influence of environment on Jewish stature. The average height of English Jewry was 170.8 cm, and Jacobs concluded that this was solely "the result of better nurture."

77. According to Maurice Fishberg, there was another cause of the consistently poor figures for Jewish stature and girth. The largest anthropometric survey of Jews had been undertaken by Snigerew in Russia and Poland for 1875 to 1877. Because of the fear of army service, Jewish males often presented themselves for national service at well below the required age of twenty, some as young as fifteen. The underdeveloped boy would then be measured, found wanting, and rejected by the army. See Fishberg, *The Jews,* 29.

78. Jacobs, *Jewish Statistics,* xii–xiii.

79. Ibid., xxiii–xxiv.

80. Renan, *Le Judaisme,* 25.

81. Jacobs, *Jewish Statistics,* xxiii–xxiv.

82. See Boas, "Changes in Bodily Form of Descendants of Immigrants."

83. See von Luschan, "Die anthropologische Stellung der Juden"; and Lombroso, *Der Antisemitismus und die Juden,* app. 3, pp. 108–114.

84. Ripley, "Racial Geography of Europe," 343.

85. Ibid., 342.

86. Jacobs, "Are Jews Jews?," 502–503.

87. Ibid., 507.

88. See the entry "Craniometry," written by Maurice Fishberg and edited by Joseph Jacobs, in *Jewish Encyclopaedia*, 4:336.

89. Jacobs had a long and fruitful relationship with Galton. In 1885 they photographed a number of boys at the Jews' Free School in London, with the aim of determining with scientific accuracy the "Jewish expression." The boys were selected upon being declared by the two researchers to have displayed a typical Jewish appearance. For the results, see Jacobs's "Types," in *Jewish Encyclopaedia*, 12:294. One year later, in 1886, the two collaborated on Jacobs's "Experiments on Prehension," "the mind's power of taking on certain material, in this case auditory sensations." See Pearson, Francis Galton, 2:272. Galton also worked on Jewish anthropology independently of Jacobs, especially in the context of his research on fingerprints. These led him initially to note "a distinct statistical difference between the finger markings of the Hebrew and the Anglo-Saxon." He later expressed doubt on this point. See Pearson, *Francis Galton*, 3B:485. As the inventor of the term "eugenics," Galton (and other eugenists) singled out Jews as the most eugenically minded and well selected of peoples. See the interview with him entitled "Eugenics and the Jew," 16.

90. Jacobs, "The Comparative Distribution of Jewish Ability," in *Jewish Statistics*, li. Some years later, in 1899, still concerned with comparative abilities and achievements, Jacobs wrote to Galton requesting his advice in choosing the correct statistical methodology for "concocting a sort of economic dummy, which I intend to call the 'Mean Englishman' . . . which I think might be a useful sociological guage for comparative purposes between period and period, and nation and nation." Jacobs to Galton, March 10, 1899, Galton Papers, no. 266, University College London. In 1919, Jacobs's *Jews of Distinction: 1815–1915* was published posthumously. It was a grand list "of those Jews who have obtained distinction in various branches of life during the past hundred years. . . . They are selected from lists numbering over nine thousand Jewish names . . . selected from 325,000 names found in the reference books of biography, etc." Jacobs, *Jews of Distinction*, introduction.

91. *Jewish Encyclopaedia*, vol. 1, preface.

92. See Israel Zangwill's comments on the occassion of a dinner held in honor of Jacobs's imminent departure for his first lecture series in the United States in 1896: *A Jewish Scholar's Career*, 4.

93. In this last position, Jacobs maintained close contact with the Büreau für Statistik der Juden in Berlin, contributing to its important journal of the same name. The Berlin enterprise is the focus of a separate section in chapter 6.

94. Jacobs, *Jewish Contributions*, 11.

95. Jacobs, *Jewish Statistics*, ii.

Five: Samuel Weissenberg: Jews, Race, and Culture

1. On the significance of the transformation that took place in the way science was presented and its rapidly increasing alienation from the mass market, see Markus, "Why Is There No Hermeneutics of Natural Sciences?"

2. In turn-of-the-century Germany and Austria, where antisemitism was a visible feature of the political and cultural landscape, it may well be that with the professionalization of scientific racism, and the subsequent closure of popular accessibility to such literature, the path was cleared for the emergence of a more crude, pseudo-scientific racism that became widely available in the form of penny pamphlets. Most representative of this kind of literature that popularized race theories, combining them with Aryan Christianity and the most vicious antisemitism, was the influential *Ostara: Briefbücherei der Blonden und Mannesrechtler*. Published from 1905 to 1918, with one issue appearing in 1922, and then again from 1927 to 1930, by the Viennese occultist antisemite Jörg Lanz von Liebenfels, the journal is one of the few items that historians know with certainty was read by the young Adolf Hitler. See Mosse, *Toward the Final Solution*, 99; for the influence of Lanz on Hitler, see Daim, *Der Mann, der Hitler die Ideen gab*.

3. Weissenberg, like his predecessor Jacobs, has an enormous bibliography to his credit. It was, however, far less eclectic, tending to focus on Jewish physical and cultural anthropology. Nevertheless, his corpus (which includes many book reviews) runs to about two hundred items.

4. This is the opinion of Ephraim Fischoff, the author of an encyclopedia entry on Weissenberg. *Encyclopaedia Judaica*, 16:417–418.

5. By a twist of historical fate, Weissenberg and Jacobs were not only bound to each other because of their anthropological pursuits: the pogroms that broke out in Russia in April 1881, and were broadcast to the west by Jacobs, began in Elizavetgrad.

6. *Encyclopaedia Judaica*, 10:1049.

7. One of the more bizarre attempts to inject religiosity into the community was undertaken by Jacob Gordin, the Yiddish playwright and journalist who founded his own sect, the Dukhovno-Bibliyskoye Bratstvo (Spiritual Biblical Brotherhood), in 1880. This anti-Talmudic group rejected postbiblical Judaism, claiming only to adhere to the ethical teachings of the Bible. It sought to abolish all Jewish fast days and holidays, repudiated commerce, and advocated Jewish occupational reform, especially, and predictably, the adoption of farming as a means of curing Russian antisemitism. In 1891 the czarist police disbanded the organization and Gordin made for New York, where he became a seminal figure in the development of the Yiddish theater. On Gordin and the sect's activities, see Lifschutz, "Jacob Gordin's Proposal to Establish an Agricultural Colony"; and Marmor, *Jacob Gordin* [Yiddish].

8. On the *numerus clausus*, see Dubnow, *History of the Jews in Russia and Poland*, 2:348–353; and Greenberg, *The Jews in Russia*, 2:34–35.

9. Weissenberg, "Curriculum Vitae."

10. In his dissertation, Weissenberg was only able to provide a bibliography of nine items, testimony enough to the paucity of research in the field to that time. See his *Ein Beitrag zur Lehre von den Lesestörungen.*

11. On the way the Prussian authorities and Jewish institutions dealt with the flood of Jewish refugees from czarist Russia, see Wertheimer, *Unwelcome Strangers.* For an excellent study of how Germans and German Jews viewed East European Jews and their culture, see Aschheim, *Brothers and Strangers.*

12. See Wertheimer, "The 'Ausländerfrage' at Institutions of Higher Learning"; and idem, "Between Tsar and Kaiser." For a contemporary statistical analysis of the situation, see Blau, "Die Juden auf den preussischen Universitäten." For a general discussion of antisemitism at Prussian universities in Wilhelmine Germany, see the thorough two-part study by Kampe, "Jews and Anti-Semites at Universities in Imperial Germany."

13. Weissenberg, "Der jüdische Typus," 309.

14. Weissenberg, "Die südrussischen Juden," 347–350.

15. In a footnote to this theory Jellinek remarked, "Obwohl jede physiologische Vergleichung mir fern liegt, so möchte ich doch hervorheben, dass Basstimmen unter den Juden viel seltener als Baritone sind." *Der Jüdische Stamm,* 89–90.

16. These were birth; the end of the first teething period; the end of the second teething period; and adulthood. Weissenberg, "Die südrussischen Juden," 349–354.

17. Ibid., 361.

18. Jacobs, "On the Racial Characteristics of Modern Jews," xxiii.

19. Weissenberg, "Die südrussischen Juden," 362.

20. Weissenberg, "Die südrussischen Juden," 368.

21. For background on the situation as it unfolded in Germany, see Fischer, *Judentum, Staat und Heer.* Just one year after Weissenberg was writing, the Comite zur Abwehr antisemitischer Angriffe in Berlin published, as part of an apologetic series defending German Jewry against antisemitic attacks, *Die Juden als Soldaten* (1896), a staunch defense of the Jew as a soldier. Presented in the form of lists, the book details the participation of and decorations awarded to Jewish soldiers in the wars of 1813–15, 1848–50, 1864, 1866, and 1870–71. It also contains an appendix of similar information for Jews in other European countries and the United States.

22. Universal military service was introduced in Russia in 1874 for all male citizens twenty-one years of age.

23. For a comparison of the size of Jews and other Europeans, see Stieda, "Ein Beitrag zur Anthropologie der Juden."

24. Quoted in Irving Howe, *World of Our Fathers* (New York, 1976), 6–7.

25. See Fishberg, *The Jews,* 86.

26. Weissenberg, "Die südrussischen Juden," 373.

27. Ibid., 375.
28. Ibid., 389 and 391.
29. Ibid., 410.
30. ". . . und betrachtet man die Juden vom rein anthropologischen Standpunkte, so ist es nicht immer so leicht, einen Juden von einem Nichtjuden zu trennen." Ibid., 562.
31. Quoted in Poliakov, *History of Anti-Semitism,* 4:87.
32. Weissenberg, "Die südrussischen Juden," 562 and 564.
33. The seven Jewish types were classified as "course," "fine," "Slavic," "southern European," "northern European," "general Caucasian," and "Mongoloid." Ibid., 568–573.
34. Ibid., 574.
35. Ibid., 575.
36. Ibid., 577. On the Khazars and their conversion to Judaism, see Dunlop, *History of the Jewish Khazars.* Renan, *Le judaisme,* 23–24, even referred to the importance of the Khazar conversion for the anthropological study of the Jews of southern Russia.
37. Weissenberg was not alone in this opinion. His contemporary H. M. Baratz traced the Hebrew influence on the earliest Russian writings. As late as 1948 the distinguished Jewish historian Cecil Roth wrote of the Khazar conversion: "To the present day, the Mongoloid features common amongst the Jews of eastern Europe are, in all probability, a heritage from these 'proselytes of righteousness' of ten centuries ago." See the third edition of Roth, *Short History of the Jewish People,* 288. For a highly critical appraisal of Baratz's work, but one that still suggests "that the Jews of Khazaria at some points and in some manner influenced the rising of Russian literature," see Dunlop, "H. M. Baratz." Support for this theory comes from recent scholarship as well. Peter Golden (*Khazar Studies,* 1:17) has written: "Khazaria lived on, however, in the contributions it made to the cultures of the peoples with which it was in contact. As to be expected, it played a great role in early Kievan history." Golden, too, makes the case for the influence of Hebrew literature in Kievan Rus: "The Old Testament, of course, came to Rus with Byzantine Orthodox Christianity. But there is evidence that individual elements of the Old Testament and Apocrypha were translated directly from the Hebrew. One of the early classics (eleventh century) of Old Rus Translation Literature was the *History of the Jewish War* of Josephus Flavius (translated from the Greek). At the same time, the *Book of Yosippon* was also translated from its Hebrew original. Josephus' *History* had considerable literary influence on Old Rus Literature, contributing a number of stock phrases, often repeated in later works."
38. *Encyclopedia Judaica,* 10:950–951. The authenticity of the letters has been the subject of heated debate. For a review of the dispute, see Dunlop, *Jewish Khazars,* 116–170.
39. The Caucasus had for some time been an important center of anthropo-

logical research. See von Erckert, *Der Kaukasus und seine Völker,* 301, where he had already anticipated by some years Weissenberg's contention that the current brachycephalism of East European Jews had its origins in the intermixture of dolichocephalic Jews with roundheaded Caucasians. Idem., "Kopfmessungen kaukasischer Völker"; and Weinberg, "Die transkaukasischen Juden."

40. Weissenberg, "Die südrussischen Juden," 578–579.

41. See von Erckert, *Der Kaukasus,* 299.

42. See Weissenberg's programmatic essay "Jüdische Museen und jüdisches in Museen."

43. On the historical background to the establishment of Jewish folklore studies in Germany and the founder of the movement, see Daxelmüller, "Max Grunwald." A brief appraisal of Grunwald's work was written by Dov Noy, the Max Grunwald Professor of Jewish Folklore at the Hebrew University. See his "Dr. Max Grunwald—the Founder of Jewish Folkloristics." For a broader survey of the field of Jewish folkloristics from Grunwald and to the present time, including a consideration of how the priorities of scholars have either been modified or changed altogether over time, see Noy, "Eighty Years of Jewish Folkloristics."

44. Weissenberg noted that it was sometimes hard to tell a Christian amulet from a Jewish one. Even more startling was his claim that "the Jews often wear Christian amulets and the Christians Jewish ones." He gave as a specific example the following description of an amulet: "A little piece of paper with the inscription: Jesus was born, Jesus was baptized, Jesus was resurrected. This amulet against malaria was prescribed by a local watchmaker and was carried by Christians and Jews." It goes without saying that this particular amulet was probably not widely worn among Jews. Nevertheless, in Weissenberg's descriptions of twenty-three separate amulets that he had collected, many, though not as overtly Christian as the one described above, were worn by Jews which were clearly pagan in origin. Weissenberg, "Südrussische Amulette."

45. Quoted in Aschheim, *Brothers and Strangers,* 39.

46. Weissenberg, "Beiträge zur Volkskunde der Juden," 130.

47. Ibid.

48. Ibid.

49. Weissenberg, "Jüdische Sprichwörter."

50. See Spector, ed., *Der Hoyzfraynd,* 1–49. This initial work of Bernstein's was greatly expanded, and he eventually gathered 3,993 different proverbs, which he published in 1908 in a magnificent volume entitled *Jüdische Sprichwörter und Redensarten.* Weissenberg wrote a glowing review of this book, calling it "a lasting monument" to the field, claiming that its value was sure to extend beyond Jewish circles to all folklorists. *Globus* 93, no. 15 (1908): 242. Bernstein supplemented this great achievement with a separate volume of 227 Yiddish proverbs about sex, which he also published in 1908: *Jüdische Sprichwörter und Redensarten: Erotica und Rustica.*

51. On the relatively late survival of Yiddish in Germany, see Freimark, "Lan-

guage Behaviour and Assimilation"; and Lowenstein, "The Yiddish Written Word."

52. Weissenberg, "Jüdische Kunst und jüdisches Kult- und Hausgerät," 202.

53. Ibid.

54. On the particular appropriation of *völkisch* culture by Jews, with its emphasis on rebirth and regeneration based on a rootedness in the past, see Mosse, "The Influence of the Volkish Idea on German Jewry."

55. Among these works are "Kinderfreud und -leid bei den südrussischen Juden" (1903); "Die Karäer der Krim" (1903); "Das Purimspiel von Ahasverus und Esther" (1904); "Die Fest- und Fasttage der südrussischen Juden" (1905); "Eine jüdische Hochzeit in Südrussland" (1905); "Speise und Gebäck bei den südrussischen Juden" (1906); "Das Feld- und Kejwermessen" (1906); "Krankheit und Tod bei den südrussischen Juden" (1907); "Palästina in Brauch und Glauben der heutigen Juden" (1907); "Das neugeborene Kind bei den südrussischen Juden" (1908). In the introduction to "Speise and Gebäck," Weissenberg again asserted the link between Jewish culture and physical existence. Jewish dietary laws had, at least since the emancipation debates in eighteenth-century France and Germany, been seen as a hindrance to full Jewish integration because they forbid observant Jews from partaking of the same food as non-Jews. Arguing in a circular fashion, he praised the segregating nature of Jewish dietary laws when he wrote: "The Jewish dietary laws belong to those factors which have given rise to the separation, and thereby the preservation, of the Jews. The separate table led to a separate existence, the separate existence to various groundless accusations, the accusations to persecution, which transformed the purely external separation into a necessary and [culture-preserving] seclusion."

56. Weissenberg, "Das Purimspiel," 1–2. Weissenberg's point about the limited use of Hebrew directly contradicts his earlier optimistic remarks about its rebirth and colloquial use in Eastern Europe. But this was an extremely polemical piece, and it appears that he has in general exaggerated the facts about the celebration of Purim in southern Russia.

57. For example, in 1905 Weissenberg complained that among all the anthropological works on the Jews that had appeared in the last decade, only one had dealt specifically with Sephardim. He sought to redress this lacuna. See Weissenberg, "Das jüdische Rassenproblem."

58. See the report of Weissenberg's journey, "Bericht über den Stand der Rudolf Virchowstiftung für 1908."

59. On the ideological imperatives behind Jewish physical anthropology's positive evaluation of the Sephardim as against the Ashkenazim, see Efron, "Scientific Racism."

60. The quotation appears in Weissenberg, "Das jüdische Rassenproblem," 6. See also von Luschan, "Die anthropologische Stellung der Juden"; and idem, "Zur physischen Anthropologie der Juden."

61. Before setting out for Palestine, Weissenberg thought that he had engi-

neered a novel experiment at home to trace the Jewish *Urtypus*. "I assume that it is well known that among contemporary Jews, Aaronides (Kohanim) and Levites are to be met with who, in the life of today's Orthodox communities, play a certain religious role and enjoy some privileges. Some of these families keep a centuries-old album and seek to marry only with irreproachable families. Now it appears to me that an anthropological study of these supposed descendants of the ancient Levites and priests could provide some information on the condition of the *jüdischer Urtypus*." Weissenberg was mistaken. The Aaronides and Levites were as short-headed and wide-faced as any Jews; like Jews in general, blondes among them amounted to 10 percent of the population, and finally, the noses of the two groups were "equally Semitic." Weissenberg concluded, "This short study shows that considered anthropologically, the Aaronides and Levites represent, on the whole, the same type as the [common] Jews. From these results it would be fundamentally incorrect to draw the conclusion that today's East European Jews are direct descendants of the ancient Israelites. For who can guarantee or bring proof that today's Kohanim are descendants of the one-time Priests? . . . The anthropometric method can discover no difference between these two groups and the Jews." Weissenberg, "Beitrag zur Anthropologie der Juden." A summary of the work appeared in *ZDSJ* 4, no. 3 (1908): 48.

62. Weissenberg, "Die autochthone Bevölkerung Palästinas."
63. Ibid., 131.
64. The Harvard anthropologist Henry Minor Huxley examined thirty-five adult male Samaritans from an anthropometric point of view in 1901. Accepting von Luschan's theory of Jewish descent, Huxley claimed that the Samaritans, with their Jewish physiognomy, were derived from the same Hittite, Amorite, and Semitic blend. Due to their close inbreeding, Huxley concluded that "the Samaritans have thus preserved the ancient type in its purity; and they are to-day the sole, though degenerate, representatives of the ancient Hebrews." See Huxley's entry on Samaritans in the *Jewish Encyclopaedia*, 10:674–676. A German version of this work appeared in *ZDSJ* 2, nos. 8 and 9 (1906): 137–139.
65. Weissenberg, "Die autochthone Bevölkerung," 137. This does not mean that Weissenberg was about to exaggerate the "purity" of the indigenous Jewish population of Palestine. His description of his visit to Peki'in, in the northern Galilee, containing his valuable description of early Jewish colonization efforts and the conditions to be found in the country at the time, brought him into contact with a Jewish population that he declared to be of mixed race. See also Weissenberg, "Peki'in und seine Juden."
66. A useful summary of the anthropological work done on the Jews of Yemen up to the 1930s is Brauer, *Ethnologie der jemenitischen Juden*.
67. For the former view, see von Maltzan, *Reise in Südarabien*, 175. In contrast to this, see Burchardt, "Die Juden in Jemen."
68. Weissenberg, "Die jemenitischen Juden," 321.
69. Weissenberg, "Die kaukasischen Juden."

70. Weissenberg, "Die zentralasiatischen Juden."
71. Weissenberg, "Zur Anthropologie der persischen Juden."
72. Weissenberg, "Die zentralasiatischen Juden," 106. In 1912 Weissenberg traveled to Central Asia, the Caucasus, and Persia in order to examine in their place of birth those same groups of Jews he had studied in Jerusalem in 1908. Not surprisingly, most of his earlier results were confirmed. For example, a trip to Turkestan to study more Central Asian Jews led him to conclude that "today's Central Asian Jews are widely removed from their *Urtypus.*" See Weissenberg, "Die zentralasiatischen Juden."
73. Weissenberg, "Die persischen Juden."
74. Weissenberg, "Die Spaniolen," 235.
75. Weissenberg also studied a group of forty-nine Jews from Urfa, Kurdistan, and Baghdad who he believed had remained a "pure remnant" (*unverfälschte Reste*) of the dispersed Jewish colonies that arose during the Babylonian exile. Weissenberg now assumed that it was possible that the Jews of Kurdistan could be descendants of those Jews who were taken into captivity at the time of the war against Nebuchadnezzar, thus forming part of the Lost Tribes. Weissenberg, "Die mesopotamischen Juden."
76. Weissenberg, "Die syrischen Juden."
77. Weissenberg, "Die Spaniolen," 235.
78. Ibid., 234.
79. Weissenberg, "Zur Anthropologie der nordafrikanischen Juden."
80. Weissenberg, "Armenier und Juden." For the conflicting opinion, see von Luschan, "Die Tachtadschy." Concerning the Palestinian Jews who had occasion to mix with these Hittites, he noted "that they were not all originally Semites," 50; and idem, "The Early Inhabitants of Western Asia.". Von Luschan's theory was supported some years before by the Jewish Viennese physician Leo Sofer. See his "Armenier und Juden."
81. On the link between eugenics and race science, see Stepan, *Idea of Race*, 111–139. For Germany, see Weindling, *Health, Race and German Politics*; and on the growth of the eugenics movements in England and the United States, see Kevles, *In the Name of Eugenics*.
82. Weissenberg, "Die Formen des ehelichen Geschlechtsverkehrs." In discussing the modes of contraception among Jews as compared with Russians, Weissenberg concluded that although the lower classes in Russia made widespread use of contraception, the Jews used various forms of birth control at twice that rate. Also, their use was to be found among all age groups among the Jewish population, while among Russians it was confined to the relatively young. Ibid., 616; idem, "Zur Biotik der südrussischen Juden." Not only had Weissenberg portrayed East European Jewry as the cultural savior of West European Jewry, but in this article he saw them as the biological savior as well. In complete accord with the Berlin sociologist Arthur Ruppin, Weissenberg held that were it not for the constant infusion of East European blood, "the liquidation of West European Jewry

would proceed at a far quicker tempo." Ibid., 200. Weissenberg made the comment in the context of his positive response to Felix Theilhaber's recently published *Der Untergang der deutschen Juden* (1911). On Theilhaber, see chap. 6 below; Weissenberg, "Beiträge zur Frauenbiologie"; idem, "Zur Sozialbiologie und Sozialhygiene der Juden."

83. Even in the midst of his intense travel through Central Asia in 1912, Weissenberg continued to publish ethnographic papers on different aspects of Jewish religious and cultural life. See, for example, the following articles: "Das Sukkothfest in Südrussland" (1912); "Weihnachtskerzen—Chanukahkerzen" (1912); "Jüdischer Volkskalender" (1913); "Rothschild-Legenden" (1913); and "Alte jüdische Grabdenkmäler aus der Krim" (1913).

84. Even on the question of women's work, he wrote: "I am not ashamed [to say]—and all gynecologists are in agreement on this—work for the woman can only be of harm." He qualified this remark by saying that he was referring only to married women and excessive work. See Weissenberg, "Zur Sozialbiologie," 408.

85. Ibid., 416.

86. Weissenberg, "Zur Biotik," 205.

87. Weissenberg, "Zur Sozialbiologie," 416.

88. Weissenberg, "Palästina in Brauch," 261.

Six: Zionism and Racial Anthropology

1. Sandler, *Anthropologie und Zionismus*, 51.

2. Kautsky, *Are the Jews a Race?*, 17.

3. Of course, these concerns weighed heavily on many sectors of society, including political Zionists. A general summary of the use of the race concept by German Zionists and how they thought its application would assist in the battle against antisemitism and assimilation is in Doron, "Rassenbewusstsein und naturwissenschaftliches Denken."

4. Mosse, *German Jews beyond Judaism*.

5. For a fuller background to the social conditions under which German Jewry lived prior to full emancipation, see Toury, *Soziale und politische Geschichte der Juden*.

6. A general discussion of how the organized German Jewish community dealt with racial antisemitism is in Bacharach, "Jewish Confrontation with Racist Anti-Semitism."

7. On Zionist interpretations of Jewish history and the development of a Zionist historical consciousness, see Almog, *Zionism and History*.

8. In its infant stages, the German Zionist movement envisioned the creation of a Jewish national homeland in Eretz Yisrael to be primarily for East European Jewry. Max Bodenheimer, a founder of the Nationaljüdische Vereinigung für Deutschland, in his pamphlet "Wohin mit den russischen Juden?" (1891) ex-

pressed the philanthropic strain of German Zionism that primarily saw the future Jewish state as a haven for the suffering masses of Russian Jews, supported by their more fortunate (and sedentary) brethren in the West. Poppel, *Zionism in Germany,* 22. The formal abandonment of this position came with the resolution adopted at the 1912 Delegiertentag held at Posen, when the ZVfD declared that it was "the obligation of every Zionist—above all the financially independent ones—to incorporate emigration to Palestine in his life program." Ibid., 50.

9. Herzl, *The Jewish State,* 33.

10. Quoted in Mosse, *Germans and Jews,* 78.

11. As to how this ignorance of things Jewish was played out among the German Zionists, no more telling example can be offered than the clash that took place in 1901 at the Fifth Zionist Congress between Herzl and a splinter group called the Demokratische Fraktion. Led by Leo Motzkin, Chaim Weizmann, and Martin Buber, the group deplored Herzl's concentration on diplomatic activities to the exclusion of cultural ones. To their minds, "national cultural activity is an immediate consequence of the concept of Zionism and is an indispensable component of Zionist work." Consequently, the group made the first plans for the establishment of a Jewish university in Eretz Yisrael and in 1902 established the Zionist publishing house Jüdischer Verlag. The full intentions of the Fraktion were spelled out in their *Programm und Organisations-Statut der Demokratischen Zionistischen Fraktion* (1902). On the political maneuvering and acrimonious debate at the Fifth Zionist Congress, see the detailed description of events in Klausner, *Opositsia le-Herzl,* 141–148. The Fraktion had obviously been greatly inspired by the theorist of cultural Zionism Ahad Ha'Am, who as early as 1889 published his celebrated essay "Lo zeh ha-derech" ("That Is Not the Way"), in which he claimed that neither colonization nor the establishment of a Jewish state indicated a Jewish national rebirth. Rather, a renaissance of Jewish education and a program of spiritual, cultural, and ethical renewal were fundamental to future colonization plans. Recently Michael Berkowitz has suggested that the split between the political and cultural Zionists was not as wide as historians have generally portrayed it to have been. See his *Zionist Culture and West European Jewry,* 40–76.

12. See Max Jungmann's rhetorically titled article "Ist das jüdisches Volk degeneriert?" (1902). Jungmann, the publisher of the satirical Zionist periodical *Schlemihl,* was an extreme Gobineauian who approvingly quoted the French thinker's axiom that degeneration referred to a people in whose veins there no longer flowed the blood of the fathers as a consequence of various "intermixtures and crossings." For Jungmann, who considered the Jews a pure race, there was no danger of them being degenerate.

13. A significant Zionist elaboration of this idea was made by the Zionist thinker and liberal Jew Franz Oppenheimer. See his controversial 1910 article, which sums up the attitude of pre–World War I German Zionism, "Stammesbewusstsein und Volksbewusstsein."

14. On the battle waged between the two representative bodies of German

Jewish liberalism and Zionism, the CV and the ZVfD, and the fundamental differences in their respective worldviews, see Reinharz, *Fatherland or Promised Land*. For an excellent history of the CV and its battle against both antisemites and Zionists, see Schorsch, *Jewish Reactions to German Anti-Semitism*.

15. There is absolutely no question as to the CV's dogged opposition to conversion. Felix Goldmann, an influential member of the CV, stated unequivocally: "The main criterion [for remaining a Jew] is, and shall remain, religion. Whoever abandons our religion is an assimilationist, whoever remains true to this religion with all his heart, may, without fear of damaging the Jewish tradition, assimilate himself culturally to his heart's content." Quoted in Reinharz, *Fatherland or Promised Land*, 83.

16. On the goals of the Wissenschaft des Judentums, see the programmatic essay by one of the founders of the Verein für Kultur und Wissenschaft der Juden, Immanuel Wolf, "On the Concept of a Science of Judaism (1822)."

17. Page references in parentheses in the text are to Auerbach, "Die jüdische Rassenfrage."

18. These included the noted physiologists Hans Friedenthal, Paul Rona, and Ulrich Friedemann, a cousin of Herzl's biographer Adolf Friedemann. The most celebrated of Auerbach's professors was the pioneer of cerebral physiology Hermann Munk, who was nominated in 1896 to occupy the chair of the late Emil Du Bois-Reymond but was rejected when he refused to submit to baptism. Auerbach, however, did not confine himself to his medical studies. Displaying a voracious appetite for learning, he also attended lectures in various other disciplines. Auerbach recalled that his first two years at Berlin "broadened his intellectual horizon": he attended lectures on ethics by Georg Simmel, national economy by Adolph Wagner, Gustav von Schmoller, and Franz Oppenheimer, criminal law by Franz von Liszt, and the history of literature and art. See Auerbach, *Pionier der Verwirklichung*, 126.

19. Auerbach's primary works in these fields were a novel, *Joab, ein Heldenleben* (1920); *Die Prophetie* (1920), a psychological examination into the nature of prophecy based on the religious experience of Jeremiah; and *Wüste und Gelobtes Land* (1938), a history of Israel from its beginning until the return from Babylonian exile. The author Arnold Zweig heartily recommended to Sigmund Freud that he read this book in preparation for his own study of Moses. In his letter to Freud, Zweig somewhat sarcastically noted: "There is just one step Au. does not take: he does not turn Moscheh into Mizri [an Egyptian]. It makes extremely stimulating, exciting reading and is written with exemplary objectivity. ... Nothing mythological, just an attempt to write history" (March 1, 1935). Freud took Zweig's advice and read Auerbach; he then smugly replied, "I was disappointed in my expectations and confirmed in my opinion. My revolutionary new ideas are not dreamt of by A." (March 14, 1935). See Ernst L. Freud, ed., *Letters of Freud and Zweig*, 103–105. It is fitting that Auerbach's final work of history was his own biography of the Jewish leader, *Moses* (1953, Eng. 1975).

Auerbach also achieved prominence as a Bible translator. In 1924 Leo Baeck headed a project to produce a new English translation of the Bible, under the editorship of Harry Torczyner. Auerbach shared with the editor the heaviest load and was personally responsible for translating Joshua, Jona, Haggai, Esther, Daniel, Ezra, Nehemiah, and Chronicles. On the project, see Ben-Chorin, "Jüdische Bibelübersetzungen in Deutschland."

20. Writing of his traditional upbringing and eventual encounter with the natural sciences, Auerbach noted, "I knew of no other Judaism than that of my father, whom we all adored and who was a model of upright living. My first glimpse beyond this narrow [Orthodox] circle came about through my early reading of natural science. However, this did not yet create a conflict between those two systems of thought—the religious and the scientific." Auerbach, *Pionier der Verwirklichung,* 98.

21. Walk, *Kurzbiographien zur Geschichte der Juden,* 13.

22. Quoted in *Encyclopaedia Judaica* (1971), 15:134.

23. In "The Origin of the Jewish Genius," Part III of *The Jews and Modern Capitalism,* Sombart included a discussion of "the race problem." A cursory and jejune survey of the current anthropological literature on the Jews led him to conclude: "For more than two thousand years [the Jews] have been untouched by other peoples; they have remained ethnically pure. That drops of alien blood came into the Jewish body corporate through the long centuries of their dispersion no one will deny. But, so small have these outside elements been that they have not influenced to any appreciable degree the ethnical purity of the Jewish people." This is, in a nutshell, the fundamental view of Auerbach. See Sombart, *The Jews and Modern Capitalism,* 285.

24. See Fishberg, *The Jews,* 191. For a contrary view, see Elkind, "Versuch einer anthropologischen Parallele zwischen den Juden und Nichtjuden."

25. *Song of Songs Rabbah* 1:1, 10.

26. On Jewish proselytism, see Graetz, *Die jüdischen Proselyten im Römerreiche;* Braude, *Jewish Proselyting;* Rosenbloom, *Conversion to Judaism;* Eichhorn, ed., *Conversion to Judaism;* Wacholder, "Attitudes towards Proselytizing in the Classical Halakha"; idem, "Cases of Proselytizing in the Tosafist Responsa"; Golb, "Conversion of European Christians"; and Seltzer, "Joining the Jewish People." On anti-Judaism in the thirteenth century, see Cohen, *The Jews and the Friars.*

27. Auerbach, "Die jüdische Rassenfrage," 337. In all matters Jewish, Auerbach was deeply influenced by the towering personality of his brother-in-law, Heinrich Loewe, whose views on gentile conversion to Judaism were clearly reflected in Auerbach's thesis. The two men were in almost total agreement that no appreciable Jewish proselytism had occurred and hence conversion to Judaism certainly had not had a physiological impact on the Jewish people. Only contemporary assimilation posed a threat to Judaism. Furthermore, Loewe was confident that Jewish racial purity was unlikely to be compromised in the future.

The offspring of mixed marriages, even if they grew up to marry Jews, would not bring about a transformation of the Jewish type. Thus it was not the racial Jew but the cultural Jew who stood at the precipice of extinction. Of the huge number of mixed marriages in Germany, Scandinavia, and Hungary, Loewe wrote: "The possibility of physical changes [through mixed marriages] is not as great as the danger of spiritual dissolution [which such marriages cause]." Of proselytism and its biological significance, he concluded: "The proselytes have not hitherto visibly influenced the Jewish racial type. If a racial mixture has occurred, it was in prehistoric times, and Virchow's thesis, namely, that the Jews form a national race, is not undermined by the fact of conversion to Judaism." Loewe, *Proselyten*, 26–27.

28. On how intermarriage related to the Jewish racial question and its effects on the sociology of Jewish life, see Ruppin's important article, "Die Mischehe."

29. Quoted in Almog, *Zionism and History*, 50.

30. Auerbach, "Ideale Werte," 68, signed "Sanctus."

31. Almog, *Zionism and History*, 23.

32. Quoted in Zimmermann, "Jewish Nationalism and Zionism in German-Jewish Students' Organizations," 146.

33. Auerbach, "Zur Verwahrung," 35.

34. Virchow, "Die Mumien der Könige"; Flinders Petrie, "The Earliest Racial Portraits," 128.

35. See Mayr, "Die bayerische Jugend"; Schimmer, "Erhebungen über die Farbe"; and Virchow, "Gesamtbericht."

36. Almost every anthropologist dealt with this perplexing issue. See Beddoe, "On the Physical Characteristics of the Jews"; Jacobs and Spielmann, "On the Racial Characteristics of Modern Jews"; Lagneau, "Sur la race juive et sa pathologie"; Ripley, *The Races of Europe*, 368–400; Judt, *Die Juden als Rasse*, 43–59; Wateff, "Anthropologische Beobachtungen der farbe"; "Hair," in *Jewish Encyclopaedia*, 6:157–160; and Fishberg, "Probleme der Anthropologie der Juden." The large Russian literature on the subject is summarized in Fishberg, *Materials for the Physical Anthropology of the Eastern European Jews*, 1:130–134.

37. Auerbach, *Palästina als Judenland*, 47.

38. Ibid., 24.

39. Von Luschan, "Offener Brief an Herrn Dr. Elias Auerbach."

40. Ibid., 370–371.

41. Ibid., 371.

42. Auerbach, *Pionier der Verwirklichung*, 187.

43. Page references to *Der Untergang der deutschen Juden* appear in parentheses in the text.

44. I wish to express my gratitude to Felix Theilhaber's son, Mr. Adin Talbar of Jerusalem. During a number of lengthy interviews in October and December of 1989, he provided me with a wealth of personal reminiscences about his fath-

er's life and work, as well as much information on the problems confronting other German Jewish emigré physicians who made their way to Palestine in the 1930s. I am particularly grateful for Mr. Talbar's kindness in lending me precious and rare copies of his father's works.

45. Heuer, "Der Untergang der deutschen Juden. Felix A. Theilhabers Darstellung der jüdisch-deutschen Identitätsproblematik" in Altenhofer and Heuer, eds., *Probleme deutsch-jüdischer Identität,*, 1:76–77.

46. Theilhaber's campaign for the repeal of paragraph 218 was, on occasion, highly visible. In 1925 he led a demonstration outside the Berlin town hall, at which he protested against the abortion and contraception laws and criticized the recent Leipzig conference of the German physicians' organizations, which, while calling for lighter sentences for those performing and having illegal abortions, betrayed their truly conservative opinions by referring to the "crime of abortion" and the "plague of abortion." On the issue of abortion during the Weimar Republic, see Atina Grossman, "Abortion and Economic Crisis: The 1931 Campaign Against Paragraph 218," in Bridenthal, Grossman, and Kaplan, eds., *When Biology Became Destiny,* 66–68. One result of Theilhaber's demonstration was the publication of a series of monographs collected under the title, *Beiträge zum Sexualproblem,* to which Theilhaber contributed five of the nineteen articles. The last of these, *Blutwunder und Liebeswahn,* concerned the celebrated stigmata case of the Bavarian farmer Therese von Konnersreuth. Theilhaber dismissed the condition as nothing more than the dermatological problem of a highly neurotic young woman. The front cover of the brochure, a surrealist drawing of a crucified priest and a bound penis, by the artist Willi Faber, was declared by a Berlin court to be pornographic and was confiscated.

47. On Theilhaber's work as a sexologist, see Hans Lehfeldt, "Felix A. Theilhaber—Pionier Sexologe," in Altenhofer and Heuer, *Probleme deutsch-jüdischer Identität,* 85–93.

48. Theilhaber, "Zur Lehre von dem Zusammenhang der sozialen Stellung und der Rasse mit der Entstehung der Uteruscarcinome." On cancer and its occurrence among the Jews, see Gutmann, *Rasse- und Krankheitsfrage der Juden,* 50–51.

49. Such literature was dismissed by its critics as pessimistic and alarmist. One Jewish newspaper summed up the exponents of this genre in the following way: "They all make an appeal to statistics which record a decrease in Jewish births and a constant increase in mixed marriages and conversions, etc. [These figures] require that the most prompt salvation [of the Jews] from their death throes is necessary. Now this literature, if one may be permitted to call the research of Rathenau, Fromer, Kohler, et al. *Auflösungsliteratur,* has enriched itself with a new statistical source book. For Zollschan, who understands the language of numbers quite well, has now been followed by Dr. Felix Theilhaber, who, with a weighty mathematical apparatus, seeks to prove the impending demise of German Jewry." See *Der Israelit,* no. 32 (1911): 1.

50. Theilhaber, "Der Untergang der deutschen Juden," *Jüdisches Literatur-Blatt* 33 (1911): 185–188.

51. *Im deutschen Reich*, no. 12 (1911): 668.

52. Theilhaber, in *Jüdische Bevölkerungs-Politik*, 8–21.

53. *Der Israelit*, no. 34 (1911): 2.

54. *Israelitische Wochenschrift*, no. 45 (1911): 1–4.

55. *Im deutschen Reich*, no. 9 (1911): 485–499. Segall mounted a one-man crusade against Theilhaber, publishing similar reviews in a number of other periodicals. See, for example, *Im deutschen Reich*, no. 12 (1911): 664–670; and "Der Untergang der deutschen Juden," *ZDSJ* 7, no. 11 (1911): 162–165.

56. *Ost und West* (1911): 997.

57. Many of these findings have in fact been confirmed by more recent work in the field of Jewish historical demography. The best recent collection of studies on Jewish historical demography is Ritterband. ed., *Modern Jewish Fertility*. See especially the introduction, and essays by Paula Hyman, Steven M. Lowenstein, and Alice Goldstein.

58. Graetz, *Structure of Jewish History*, 233–234.

59. See Grunwald, ed., *Die Hygiene der Juden*.

60. Lestshinsky, *Dos yidishe folk in tsifern*, 120.

61. Steven M. Lowenstein, "Limitation of Fertility" in *Mechanics of Social Change*, 65–84.

62. Fishberg, *The Jews*, 232.

63. Rathenau's remark was made in an article entitled "Höre, Israel!" ("Hear, O Israel!") which he wrote for Maximilian Harden's journal *Zukunft*. In this tasteless appeal to his fellow Jews to assimilate, Rathenau's Jewish self-hatred comes through when he directs his readers to have a good look in the mirror at "your unathletic build, your narrow shoulders, your clumsy feet, your sloppy roundish shape." Theodor Herzl wrote to Maximillian Harden that there was no need to refute Rathenau because what he said was basically true. Herzl rejected Rathenau's cure for Jewish degeneration—assimilation—while accepting the diagnosis. See Bein, et al., eds., *Theodor Herzl: Briefe und Tagebücher*, vol. 4, 205.

64. See Niewyk, *The Jews in Weimar Germany*; Barkai, "Die Juden als sozioökonomische Minderheitsgruppe." Contemporary studies that are still of great value are Marcus, *Die wirtschaftliche Krise des deutschen Juden*, and Lestshinsky, *Das wirtschaftliche Schicksal des deutschen Judentums*. For a sense of the extent to which German Jews were dependent on social assistance, see also the superbly detailed guide to all of Germany's Jewish welfare institutions, Schlesinger, ed., *Führer durch die Jüdische Wohlfahrtspflege in Deutschland*.

65. Theilhaber, in *Jüdische Bevölkerungs-Politik*, 12.

66. Ibid., 67.

67. See Weindling, *Health, Race and German Politics*, 245.

68. Theilhaber, in *Jüdische Bevölkerungs-Politik*, 69. Aron Sandler was a

vocal campaigner on behalf of his Jüdische Volkspartei for communal election reform. Sandler's activities in this area are addressed by Brenner, "Jüdische Volkspartei."

69. In fact, Theilhaber questioned the value of anthropometric study of the Jews, suggesting that comparative studies of Jewish and gentile pathologies would be more worthwhile. See his "Beiträge zur jüdischen Rassenfrage."

70. Theilhaber explored the theme of Jewish identity crisis and its relation to assimilation in his last book, *Judenschicksal*, a biographical study of eight assimilated Jews: Ferdinand Lassalle, Emin Pascha, Alfred Dreyfus, Otto Weininger, Albert Ballin, Ludwig Frank, Rosa Luxemburg, and Walter Rathenau.

71. Zollschan, *Jewish Questions*, 4–5.

72. *Encyclopaedia of Zionism and Israel*, 2:1283.

73. Wininger, *Grosse jüdische National-Biographie*, 369. In fact, Zollschan had been offered a permanent lectureship at the University of Vienna in 1911 and also one at the Hebrew University in 1928, but considered it his duty to remain in private practice and pursue X-ray research. Zollschan File, CZA A122/13.

74. Zollschan, *Revaluation of Jewish Nationalism*, 7–8.

75. The best accounts of Austrian antisemitism remain Pulzer, *Rise of Political Anti-Semitism*, 127–287; Schorske, *Fin-de-Siècle Vienna*, 116–180; and Katz, *From Prejudice to Destruction*, 281–291; Oxaal, Pollack, and Botz, eds., *Jews, Antisemitism and Culture in Vienna*. For an excellent overview of how Viennese Jews responded to the radical political change in Austria from liberalism to conservatism in the late nineteenth century, see Klein, *Jewish Origins of the Psycho-Analytic Movement*, esp. 1–32.

76. Wistrich, *Jews of Vienna*, 205–206.

77. Page references to *Das Rassenproblem* appear in parentheses in the text.

78. Bacharach, "Ignaz Zollschans 'Rassentheorie,' " 180.

79. On Chamberlain, see Field, *Evangelist of Race*, 112–117; also the brief but perceptive introduction by George L. Mosse to Chamberlain, *Foundations of the Nineteenth Century*.

80. Scheuer, *Burschenschaft und Judenfrage*; idem, *Die geschichtliche Entwicklung des deutschen Studententums in Österreich*.

81. Wistrich, *Jews of Vienna*, 59–60, 215–217, 367. On Jewish student fraternities at the University of Vienna and their battle against antisemitism and for Jewish nationalism and Zionism, see Rozenblit, "The Assertion of Identity."

82. See Beller, *Vienna and Its Jews*; and Berkley, *Vienna and Its Jews*.

83. See Oxaal and Weitzmann, "The Jews of Pre-1914 Vienna," 398. Over this time there was a percentage decrease in the number of Jews in the city. Jews made up 10 percent of the city's population in 1880, 12 percent in 1890, but only 8.6 percent in 1910. It was this long-term trend that so alarmed Theilhaber and other Jewish demographers.

84. Rozenblit, *The Jews of Vienna*, 152.

85. On the Aryan theory, see Snyder, *Race,* 58–179; Barzun, *Race,* 97–114; and Herz, *Race and Civilization,* 75–128, 189–244.

86. Quoted in Patai and Wing, *Myth of the Jewish Race,* 24–25.

87. As Hannah Arendt demonstrated, the idea that dolichocephalic invaders conquered indigenous brachycephalic peoples was first elevated to a theory of history by the Comte de Boulainvilliers, a French aristocrat of the early eighteenth century, in order to explain his nation's past. Though he focused more on the issue of class than race, he believed that France, owing to invasion and conquest, was composed of two very different kinds of peoples. "[De Boulainvilliers] interpreted the history of France as the history of two different nations of which the one, of Germanic origin, had conquered the older inhabitants, the 'Gaules,' had imposed its laws upon them, had taken their lands, and had settled down as the ruling class, the 'peerage' whose supreme rights rested upon the 'right of conquest' and the 'necessity of obedience always due to the strongest.'" Arendt, *Origins of Totalitarianism,* 162.

88. Zollschan, *Rassenproblem,* 85–96. See Nyström, "Über Formenveränderungen."

89. "Racialism," Zollschan File, CZA A122 4/1.

90. Zollschan, "The Jewish Race Problem," 398.

91. Zollschan, *Jewish Questions,* 13–14.

92. Ibid., 14–15.

93. Jacobs, "The Comparative Distribution of Jewish Ability," in *Jewish Statistics,* lvi.

94. Zollschan, *Jewish Questions,* 35.

95. Ibid., 42.

96. Zollschan, "Der Kulturwert der jüdischen Rasse," *Die Selbstwehr,* n.d., 14. Zollschan File, CZA, A122/13.

97. Zollschan, "The Jewish Race Problem," 391.

98. Ibid., 402.

99. Zollschan, *Racialism against Civilization,* 57.

100. "To the Scientists of Our Time!" Zollschan File, CZA A122/1.

101. Dr. Ziegler to Stephen Wise, December 22, 1936, Zollschan File, CZA A122/12.

102. Zollschan, *Racialism against Civilization,* 63. In December 1944 Zollschan addressed the Anglo-Palestine Club in London on the "Prospects of the Racial Doctrine." He had the full support of the chief rabbi, who encouraged the Jewish Defence Committee of the Board of Deputies to assist Zollschan with his mission. The committee rejected "the Chief Rabbi's suggestion on the ground that very few people in this country took the racial doctrine seriously, that various books had dealt with it and there was no need to do anything more about it." The secretary went on to record: "This strange ante-deluvian view reflected no credit on those who claimed to be defenders of the Jewish people." Report of Zollschan's Address, Zollschan File, CZA A122/13.

103. Zollschan, *Revaluation of Jewish Nationalism*, 8.
104. Bacharach, "Ignaz Zollschans 'Rassentheorie,' " 189.
105. Zunz, "Grundlinien zu einer künftigen Statistik der Juden." See also Jacobs, "Jüdische Volkskunde." It is significant that Jacobs' division of Jewish statistics into fourteen specialties (already elucidated in his work of 1882) became the dominant paradigm for the work of the bureau. The areas he singled out for intensive study were population, demography, anthropology, psychology, sociology, criminality, economic conditions, health, education, religion, citizen's rights, organizations, social welfare, and miscellaneous studies.
106. Barzilay, *Shlomo Yehudah Rappaport*, 14. See Zunz, *Die gottesdienstlichen Vorträge der Juden*, preface.
107. For an account of Nossig's multifarious activities on behalf of the Jewish community, including a discussion of his tragic demise and execution for treason by the Jewish Fighting Organization in the Warsaw ghetto, see Almog, "Alfred Nossig: A Reappraisal."
108. Ruppin, *Memoirs, Diaries, Letters*, 75. Ruppin's criticism is difficult to understand. Not all of the contributors were dilettantes, for they included such well-known scholars as Joseph Jacobs, Maurice Fishberg, I. M. Judt, Alfred Nossig, and Ruppin himself. For more on the founding of the bureau, see Ruppin, *Memoirs*, 74–75, and the encyclopaedia entry by Felix Theilhaber, "Statistik der Juden," *Jüdisches Lexikon*, 4:631. See also "Büros für Statistik der Juden," *Jüdisches Lexikon*, 1:1249–1250.
109. Indeed, the liberal Jewish press tended to condemn the book, while the Zionist press generally applauded its findings. Ruppin, *Tagebücher, Briefe, Erinnerungen*, 131. An exception to this is the introduction to the English-language edition of the book, written by the non-Zionist Joseph Jacobs. In commending Ruppin's work, Jacobs reaffirmed his own pioneering role in the field of Jewish sociology: "Hence, when examining, some thirty years ago, into the charges which anti-Semites brought against the Jews, I found it necessary to get as many numerical facts as possible about the Jews in their various aspects, economic, demographic, and the like, and thus started a series of investigations which have grown into a whole literature, of which this book of Dr. Ruppin's is a striking example." See Ruppin, *The Jews of Today*, xv.
110. Nossig, ed., *Jüdische Statistik*, 2.
111. The original committee of the association was composed of the following men, some of whom were leading figures in the Zionist movement: Nathan Birnbaum, Martin Buber, David Farbstein, Berthold Feiwel, Alexander Hausmann, Abraham Kastelianski, Alfred Klee, Jacob Kohann-Bernstein, Abraham Korkis, Sigmund Kornfeld, Egon Lederer, Leo Motzkin, L. Felix Pinkus, Davis Trietsch, Chaim Weizmann, and Theodor Zlocisti.
112. Nordau, *Zionistische Schriften*, 113–114.

113. Nossig, ed., *Jüdische Statistik*, 147. See also Ruppin's assessment of the Bureau's first three years, "Der Stand der Statistik der Juden," 177–178.

114. A clear example of this ecumenism is that while Ruppin, a Zionist, was the journal's editor from 1905 to 1908, he was eventually succeeded, after his emigration to Palestine, by Jacob Segall. Segall, a bitter opponent of Felix Theilhaber, was also on the staff of the Centralverein deutscher Staatsbürger jüdischen Glaubens.

115. See the discussion of the Jewish racial question in Ruppin, *The Jews in the Modern World*, 8–15.

116. Curt Michaelis, "Die jüdische Auserwählungsidee und ihre biologische Bedeutung," 3.

117. Nossig, "Die Auserwähltheit der Juden im Lichte der Biologie."

118. Cheinisse, "Die Rassenpathologie und der Alkoholismus bei den Juden,". On the defensive application of Jewish statistics, see Arthur Cohen, "Statistik und Judenfrage."

119. Wassermann, "Ist die Kriminalität der Juden Rassenkriminalität?"; and Thon, "Kriminalität der Christen und Juden in Oesterreich." See also Blau, "Die Juden als Sexualverbrecher."

❊ REFERENCES ❊

Newspapers and Periodicals

Allgemeine Zeitung des Judentums. Leipzig and Berlin.
American Jewish Historical Quarterly. New York.
Anthropologischer Anzeiger. Stuttgart.
Archiv für Anthropologie. Braunschweig.
Archiv für Rassen- und Gesellschaftsbiologie. Berlin.
Das Ausland. Stuttgart.
Beiträge zur Anthropologie Bayerns. Munich.
British History Illustrated. London.
Bulletin de l'Academie de medicine de Paris. Paris.
Bulletin de la Société d'anthropologie de Paris. Paris.
Centralvereins Zeitung: Blätter für Deutschtum und Judentum. Berlin.
Correspondenz-Blatt der deutschen Gesellschaft für Anthropologie, Ethnologie und Urgeschichte. Berlin.
Deutsche Jahrbücher für Politik und Literatur. Berlin.
Deutsches Ärzteblatt—Ärztliche Mitteilungen. Berlin.
Globus: Illustrierte Zeitschrift für Länder- und Völkerkunde. Braunschweig.
Historia Judaica. New York.
Im Deutschen Reich: Zeitschrift des Centralvereins deutscher Staatsbürger jüdischen Glaubens. Berlin.
Der Israelit. Frankfurt am Main and Mainz.
Israelitische Wochenschrift. Württemberg.
Israelitisches Familienblatt. Hamburg.

REFERENCES

Jahrbuch des Instituts für deutsche Geschichte. Tel Aviv.
Jahres-Bericht des Jüdisch-Theologischen Seminars. Breslau.
Jewish Chronicle. London.
Jewish Historical Society of England: Transactions. London.
Jewish Quarterly Review. London.
Jewish Review. London.
Jewish Social Studies. New York.
Journal and Proceedings of the Australian Jewish Historical Society. Sydney.
Journal of Contemporary History. London.
Journal of Jewish Studies. London.
Journal of Mental Science. London.
Journal of the History of Medicine and Allied Sciences. New York.
Journal of the History of the Behavioral Sciences. Brandon, Vermont.
Journal of the Royal Anthropological Institute of Great Britain and Ireland. London.
Der Jude: Eine Monatsschrift. Berlin.
Jüdische Familien-Forschung. Berlin.
Jüdische Rundschau. Berlin.
Der jüdische Student: Monatsschrift des Bundes Jüdischer Corporationen. Berlin.
Jüdischer Almanach 5670. Vienna.
Jüdisches Literatur-Blatt. Berlin.
Leo Baeck Institute Year Book. London.
Magazin der Gesellschaft naturforschender Freunde für die neuesten Entdeckungen in der gesamten Naturkunde. Berlin.
Man: The Journal of the Royal Anthropological Institute. London.
Medical Record. New York.
Medizinhistorisches Journal. Stuttgart.
Memoirs American Anthropological Association. Lancaster, Pa.
Mémoires de la Société d'Anthropologie de Paris. Paris.
Menorah Journal. New York.
Midstream. New York.
Mitteilungen der Anthropologische Gesellschaft in Wien. Vienna.
Mitteilungen der Gesellschaft für jüdische Volkskunde. Hamburg.
Monatsschrift für Geschichte und Wissenschaft des Judentums. Breslau.
Nature: A Weekly Illustrated Journal of Science. London.
OSE-Rundschau: Zeitschrift der Gesellschaft für Gesundheitsschutz der Juden. Berlin.
Ost und West. Illustrierte Monatsschrift für das gesamte Judentums. Berlin
Palästina. Berlin.
Politisch-Anthropologische Revue. Later called Politisch-Anthropologische Monatsschrift. Leipzig.
Popular Science Monthly. New York.

Proceedings of the Ninth World Congress of Jewish Studies. Jerusalem.
Revue d'Anthropologie. Paris.
St. Petersburger Medicinische Wochenschrift. St. Petersburg.
Selbstwehr: Jüdisches Volksblatt. Prague.
Sitzungs-Berichte der Berliner Akademie der Wissenschaft. Berlin.
Statistische Monatschrift. Vienna.
Studies in Eighteenth-Century Culture. Cleveland.
Studies in Zionism. Haifa.
Verhandlungen der naturforschenden Gesellschaft zu Basel. Basel.
Volk und Land: Jüdische Wochenschrift für Politik, Wirtschaft und Palästinaarbeit. Berlin.
Die Welt. Vienna and Berlin.
Wiener klinische Wochenschrift. Vienna.
Wissenschaftliche Mitteilungen aus Bosnien und der Hercegovina. Vienna.
Zeitschrift des königlich bayerischen statistischen Bureaus. Munich.
Zeitschrift für Demographie und Statistik der Juden. Berlin.
Zeitschrift für die Wissenschaft des Judenthums. Berlin.
Zeitschrift für Ethnologie. Berlin.
Zeitschrift für Morphologie und Anthropologie. Stuttgart.
Zentralblatt für Nervenheilkunde und Psychiatrie.
Zeitschrift für Sexualwissenschaft. Leipzig.

Books and Articles

Abrahamsen, David. *The Mind and Death of a Genius.* New York: Columbia University Press, 1946.
Ackerknecht, Erwin H. "German Jews, English Dissenters, French Protestants: Nineteenth-Century Pioneers of Modern Medicine and Science." In *Healing and History: Essays for George Rosen.* Edited by Charles Rosenberg, 86–96. Kent: William Dawson, 1979.
Adair, James. *Adair's History of the American Indians.* Edited by Samuel Cole Williams. New York: Argonaut Press, 1966.
Almog, Shmuel. "Alfred Nossig: A Reappraisal." *Studies in Zionism,* no. 7 (1983): 1–29.
———. "The Racial Motif in Renan's Attitude to Jews and Judaism." In *Anti-Semitism through the Ages.* Edited by Shmuel Almog, 255–278. Oxford: Pergamon Press, 1988.
———. *Zionism and History: The Rise of a New Jewish Consciousness.* New York: St. Martin's Press, 1987.
Alsberg, Moritz, *Rassenmischung im Judentum.* Hamburg: J. S. Richter, 1891.
———. *Die Abstammung des Menschen.* Cassel: Th. G. Fischer, 1902.
Altenhofer, Norbert, and Heuer, Renate, eds. *Probleme deutsch-jüdischer Identität.* Vol. 1. Frankfurt: A. & V. Woywod, 1986.

REFERENCES

Alumni Cantabrigienses, Pt. II, 1752 to 1900. Vol. 3. Cambridge: University Press, 1947.

Ammon, Otto. *Die natürliche Auslese beim Menschen.* Jena: G. Fischer, 1893.

Andree, Richard. *Zur Volkskunde der Juden.* Bielefeld: Velhagen & Klasing, 1881.

Arendt, Hannah. *The Origins of Totalitarianism.* San Diego: Harcourt Brace Jovanovich, 1973.

Aschheim, Steven. *Brothers and Strangers: The East European Jew in German and German-Jewish Consciousness, 1800–1923.* Madison: University of Wisconsin Press, 1982.

Auerbach, Elias. "Bemerkungen zu Fishberg's Theorie über die Herkunft der blonden Juden." *ZDSJ* 3, no. 6 (1907): 92–93.

———. "Ideale Werte." *Der jüdische Student* 1, no. 5 (1902): 65–68.

———. "Die jüdische Rassenfrage," *ARGB* 4, no. 3 (1907): 332–361.

———. *Palästina als Judenland.* Berlin: Jüdischer Verlag, 1912.

———. *Pionier der Verwirklichung: Ein Arzt aus Deutschland erzählt vom Beginn der zionistischen Bewegung und seiner Niederlassung in Palästina kurz nach der Jahrhundertwende.* Stuttgart: Deutsche Verlags-Anstalt, 1969.

———. "Die Sterblichkeit der Juden in Budapest 1901–1905." *ZDSJ* 4, no. 10 (1908): 145–158; and no. 11 (1908): 161–168.

———. "Zur Verwahrung." *Der jüdische Student* 1, no. 3, (1902): 33–35.

Bach, H. I. *The German Jew: A Synthesis of Judaism and Western Civilization, 1730–1930.* Oxford: Oxford University Press, 1985.

Bacharach, Walter Zwi. *Giz'anut be-sherut ha-politika: Min ha-monizm el ha-nazism.* Jerusalem: Magnes Press, 1985.

———. "Ignaz Zollschans 'Rassentheorie.'" *Jahrbuch des Instituts für deutsche Geschichte,* Beiheft 6 (1983): 179–197.

———. "Jewish Confrontation with Racist Anti-Semitism, 1879–1933." *LBIYB* 25 (1980): 197–219.

Barkai, Abraham. "Die Juden als sozio-ökonomische Minderheitsgruppe in der Weimarer Republik." In *Juden in der Weimarer Republik.* Edited by Walter Grab and Julius Schoeps, 330–346. Stuttgart: Burg Verlag, 1986.

Barkan, Elazar. "Mobilizing Scientists against Nazi Racism, 1933–1939." In *Bones, Bodies, Behavior: Essays on Biological Anthropology.* Edited by George W. Stocking, Jr., 180-205. Madison: University of Wisconsin Press, 1988.

———. *The Retreat of Scientific Racism. Changing Concepts of Race in Britain and the United States between the World Wars.* Cambridge: Cambridge University Press, 1992.

Baron, Salo W. "Ghetto and Emancipation: Shall We Revise the Traditional View?" *Menorah Journal* 14, no. 6 (1928): 515–526.

Barzilay, Isaac. "The Jew in the Literature of the Enlightenment." *Jewish Social Studies* 18 (1956): 243–261.

———. *Shlomo Yehudah Rappaport [SHIR], 1790–1867, and His Contemporar-*

ies: Some Aspects of Nineteenth Century Jewish Scholarship. Israel: Massada Press, 1969.

Barzun, Jacques. *Race: A Study in Modern Superstition*. New York: Harper Torchbook, 1965.

Baskerville, Beatrice C. *The Polish Jew: His Social and Economic Value*. New York: Macmillan, 1906.

Beadles, Cecil. "The Insane Jew." *Journal of Mental Science* 46, no. 195 (1900): 731–737.

Beddoe, John. "On the Physical Characteristics of the Jews." *Transactions of the Ethnological Society of London* 1 (1861): 222–237.

Beller, Steven. *Vienna and Its Jews, 1867–1938: A Cultural History*. Cambridge: Cambridge University Press, 1989.

Ben-Chorin, Schalom. "Jüdische Bibelübersetzungen in Deutschland." *LBIYB* 4 (1959): 311–331.

Benedikt, Moritz. "The Insane Jew: An Open Letter to Dr. C.F. Beadles." *Journal of Mental Science* 47, no. 198 (1901): 503–509.

Benjamin, David J. "Joseph Jacobs." *Journal and Proceedings of the Australian Jewish Historical Society* 3 (1949): 72–91.

Berding, Helmut. *Moderner Antisemitismus in Deutschland*. Frankfurt am Main: Suhrkamp, 1988.

"Bericht über den Stand der Rudolf Virchowstiftung für 1908." *ZE* 40 (1908): 977.

Berkley, George E. *Vienna and Its Jews: The Tragedy of Success, 1880s–1980s*. Cambridge, Mass.: Abt Books, 1988.

Berlin, Isaiah. *Against the Current: Essays in the History of Ideas*. London: Penguin Books, 1982.

Bernstein, Ignatz. *Jüdische Sprichwörter und Redensarten: Erotica und Rustica*. Warsaw: B. W. Segal, 1908.

Bertholet, Alfred. *Die Stellung der Israeliten und der Juden zu den Fremden*. Freiburg: J. C. B. Mohr, 1896.

Besser, Max. *Die Juden in der modernen Rassentheorie*. Cologne: Jüdischer Verlag, 1911.

Biddiss, Michael D. *Father of Racist Ideology: The Social and Political Thought of Count Gobineau*. New York: Weybright and Talley, 1970.

Blanckaert, Claude. "On the Origins of French Ethnology: William Edwards and the Doctrine of Race." In *Bones, Bodies, Behavior: Essays on Biological Anthropology*. Edited by George W. Stocking, Jr., 18–55. Madison: University of Wisconsin Press, 1988.

Blau, Bruno. "Die Juden als Sexualverbrecher." *Im deutschen Reich* 2 (1911).

———. "Die Juden auf den preussischen Universitäten im Jahre 1905/06." *ZDSJ* 4, no. 9 (1908): 140–141.

Blau, Bruno, ed. *Statistik der Juden. Eine Sammelschrift*. Berlin: Jüdischer Verlag, 1918.

Blechmann, Bernhard. *Ein Beitrag zur Anthropologie der Juden.* Dorpat: Wilhelm Just, 1882.

Blumenbach, Johann Friedrich. *On the Natural Varieties of Mankind.* New York: Berman Publishers, 1969.

Boas, Franz. "Changes in Bodily Form of Descendants of Immigrants." *American Anthropologist* 14 (1912): 530–562.

Boas, Franz, Maurice Fishberg, and Ellsworth Huntington. *Aryan and Semite: With Particular Reference to Nazi Racial Dogmas.* Cincinnati: B'nai B'rith, 1934.

Böhlich, Walter, ed. *Der Berliner Antisemitismusstreit.* Frankfurt am Main: Insel Verlag, 1988.

Bolt, Christine. *Victorian Attitudes to Race.* London: Routledge & Kegan Paul, 1971.

Booth, Charles, ed. *The Life and Labour of the People of London.* London: Williams & Norgate, 1889–1891.

Boudin, Jean-Christian. "Sur l'idiote et l'aliénation mentale chez les Juifs d'Allemagne." *Bulletin de la societé d'anthropologie de Paris* (1863): 386–388.

———. *Traité de Géographie et de statistique médicales et des maladies endémiques.* Vol. 2. Paris: J. B. Baillière, 1857.

Boyer, John W. *Political Radicalism in Late Imperial Vienna: Origins of the Christian Social Movement, 1848–1897.* Chicago: University of Chicago Press, 1981.

Braude, William G. *Jewish Proselyting in the First Five Centuries of the Common Era: The Age of the Tannaim and Amoraim.* Providence, R.I.: Brown University Press, 1940.

Brauer, Erich. *Ethnologie der jemenitischen Juden.* Heidelberg: Carl Winters Universitätsbuchhandlung, 1934.

Brenner, Michael. "The Jüdische Volkspartei—National-Jewish Communal Politics during the Weimar Republic." *LBIYB* 35 (1988): 219–243.

Breslauer, Bernhard. *Die Zurücksetzung der Juden an den Universitäten Deutschlands.* Berlin: B. Levy, 1911.

Bridenthal, Renate, Atina Grossman, and Marion Kaplan, eds. *When Biology Became Destiny: Women in Weimar and Nazi Germany.* New York: Monthly Review Press, 1984.

Buchanan, Claudius. *Christian Researches in Asia: With Notices of the Translation of the Scriptures into the Oriental Languages.* Boston: Samuel T. Armstrong, 1811.

Buchanan, Michelle. "Savages, Noble and Otherwise, and the French Enlightenment." *Studies in Eighteenth-Century Culture* 15 (1986): 97–109.

Burchardt, Hermann. "Die Juden in Jemen." *Ost und West* 2, no. 5 (1902): 337–342.

Burke, John G. "The Wild Man's Pedigree: Scientific Method and Racial Anthropology." In *The Wild Man Within: An Image in Western Thought from the*

Renaissance to Romanticism. Edited by Edward Dudley and Maxmillian E. Novak, 259–280. Pittsburgh: University of Pittsburgh Press, 1972.

Buschan, Georg. "Einfluß der Rasse auf die Form und Häufigkeit pathologischer Veränderungen." *Globus* 67, no. 2 (1895): 21–24, 43–47, 60–63, 76–80.

Calisch, Edward N. *The Jew in English Literature, as Author and as Subject.* Port Washington, N.Y.: Kennikat Press, 1969.

Carlebach, Esriel. *Exotische Juden.* Berlin: Welt Verlag, 1932.

Cassel, David. *Offener Brief eines Juden an Herrn Professor Dr. Virchow.* Berlin: Louis Gerschel, 1869.

Chamberlain, Houston Stewart. *Foundations of the Nineteenth Century.* 2 vols. New York: Howard Fertig, 1977.

Chamberlin, J. Edward, and Sander L. Gilman, eds. *Degeneration: The Dark Side of Progress.* New York: Columbia University Press, 1985.

Cheinisse, L. "Die Rassenpathologie und der Alkoholismus bei den Juden." *ZDSJ* 6, no. 1 (1910): 1–8.

Clifford, James. *The Predicament of Culture.* Cambridge, Mass.: Harvard University Press, 1988.

Clifford, James, and George E. Marcus, eds. *Writing Culture: The Poetics and Politics of Ethnography.* Berkeley and Los Angeles: University of California Press, 1986.

Cohen, Arthur. "Statistik und Judenfrage." *ZDSJ* 1, no. 3 (1905): 11–14.

Cohen, Jeremy. *The Jews and the Friars: The Evolution of Medieval Anti-Judaism.* Ithaca: Cornell University Press, 1982.

Cohn, Norman. *Warrant for Genocide: The Myth of Jewish World-Conspiracy and the Protocols of the Elders of Zion.* Chico, Calif.: Scholars Press, 1981.

Comité zur Abwehr antisemitischer Angriffe in Berlin. *Die Juden als Soldaten.* No. 2 in the series *Die Juden in Deutschland.* Berlin: Sigfried Cronbach, 1896.

Count, Earl W. *This Is Race: An Anthology Selected from the International Literature on the Races of Man.* New York: Henry Schuman, 1950.

Cowen, Anne, and Cowen, Roger. *Victorian Jews through British Eyes.* Oxford: Oxford University Press, 1986.

Curtin, Philip D. *The Image of Africa: British Ideas and Action, 1780–1850.* Madison: University of Wisconsin Press, 1964.

Daim, Wilfried. *Der Mann, der Hitler die Ideen gab: Von den religiösen Verirrungen eines Sektierers zum Rassenwahn des Diktators.* Munich: Isar Verlag, 1958.

Darwin, Charles. *The Descent of Man and Selection in Relation to Sex.* Vol. 1. New York: J. A. Hill, 1904.

———. *Effects of Cross and Self Fertilization in the Vegetable Kingdom.* London: J. Murray, 1876.

———. *The Variation of Animals and Plants under Domestication.* Vol. 2. New York: D. Appelton, 1876.

Daxelmüller, Christoph. "Max Grunwald and the Origin and Condition of Jewish Folklore at Hamburg." *Proceedings of the Ninth World Congress of Jewish Studies*, division D, vol. 2 (1986): 73–80.
De Quatrefages, A. "A Natural History of Man." *Popular Science Monthly* 1 (1872): 61–75.
Devay, Francis. *Du danger des mariages consanguins au point de vue sanitaire.* Paris: V. Masson, 1862.
Disraeli, Benjamin. *Coningsby or the New Generation.* Harmondsworth, Eng.: Penguin, 1983.
Doron, Joachim. "Rassenbewusstsein und naturwissenschaftliches Denken im deutschen Zionismus während der wilhelminischen Ära." *Jahrbuch des Instituts für deutsche Geschichte* 9 (1980): 389–427.
Douglas, Hugh. "Burke and Hare." *British History Illustrated* 5, no. 5 (1978–79): 32–43.
Dubnow, Simon. *History of the Jews in Russia and Poland: From the Earliest Times to the Present.* 3 vols. Philadelphia: Jewish Publication Society, 1916–20.
Dunlop, Douglas Morton. *The History of the Jewish Khazars.* Princeton: Princeton University Press, 1954.
———. "H. M. Baratz and His View of Khazar Influence on the Earliest Russian Literature, Juridical and Historical." In *Salo Wittmayer Baron Jubilee Volume.* Edited by Saul Liebermann, 1:345–367. Jerusalem: Academy for Jewish Research, 1974.
Efron, John M. "Scientific Racism and the Mystique of Sephardic Racial Superiority." *LBIYB* 38 (1993): 75–96.
———. "The 'Kaftanjude' and the 'Kaffeehausjude': Two Models of Jewish Insanity: A Discussion of Causes and Cures among German-Jewish Psychiatrists." *LBIYB* 37 (1992): 169–188.
Ehrke, Thomas Rainer. "Antisemitismus in der Medizin im Spiegel der 'Mitteilungen aus dem Verein zur Abwehr des Antisemitismus' (1891–1931)." Inaugural dissertation, Mainz, Johannes-Gutenberg-Universität, 1978.
Eichhorn, David Max, ed. *Conversion to Judaism: A History and Analysis.* N.p.: Ktav, 1965.
Eisenstadt, H.L. "Die Renaissance der jüdischen Sozialhygiene." *ARGB* 5 (1908): 707–728.
Elbogen, Ismar. "Die Bezeichnung 'Jüdische Nation.'" *MGWJ* 63 (1919): 200–208.
Elbogen, Ismar, and Eleonore Sterling. *Die Geschichte der Juden in Deutschland.* Frankfurt am Main: Athenäum, 1988.
Elkind, Arkadius. "Anthropologische Untersuchungen über die russisch-polnischen Juden und der Wert dieser Untersuchungen für die Anthropologie im allgemeinen." *ZDSJ* 2, no. 4 (1906): 49–54; no. 5 (1906): 65–69.
———. "Versuch einer anthropologischen Parallele zwischen den Juden und Nichtjuden." *ZDSJ* 4, no. 1 (1908): 1–5; no. 2 (1908): 24–29.

REFERENCES

Encyclopaedia Judaica. 10 vols. Berlin: Verlag Eschkol, 1928–1934.
Encyclopaedia Judaica. 16 vols. Jerusalem: Keter, 1971.
Encyclopaedia of Zionism and Israel. 2 vols. New York: Herzl Press and McGraw Hill, 1971.
Endelman, Todd M. "Disraeli's Jewishness Reconsidered." *Modern Judaism* 5, no. 2 (1985): 109–123.
———. *The Jews of Georgian England, 1714–1830: Tradition and Change in a Liberal Society.* Philadelphia: Jewish Publication Society, 1979.
Engländer, Martin. *Die auffallend häufigen Krankheitserscheinungen der jüdischen Rasse.* Vienna: J. L. Pollack, 1902.
Erb, Rainer. "Warum ist der Jude zum Ackerbürger nicht tauglich? Zur Geschichte eines antisemitischen Stereotyps." In *Antisemitismus und Jüdische Geschichte: Studien zu Ehren von Herbert A. Strauss.* Edited by Rainer Erb and Michael Schmidt, 99–120. Berlin: Wissenschaftlicher Autorenverlag, 1987.
Erb, Rainer, and Werner Bergmann. *Die Nachtseite der Judenemanzipation: Der Widerstand gegen die Integration der Juden in Deutschland, 1780–1860.* Berlin: Metropol, 1989.
Evans, Richard, J. *The Feminist Movement in Germany, 1894–1933.* London: Sage Publications, 1976.
Evans-Gordon, William Eden. *The Alien Immigrant.* London: W. Heinemann, 1903.
Falk, Ze'ev W. *Jewish Matrimonial Law in the Middle Ages.* Oxford: Oxford University Press, 1966.
Feist, Sigmund. *Stammeskunde der Juden: Die Jüdischen Stämme der Erde in alter und neuer Zeit.* Leipzig: Hinrichs, 1925.
Feldman, David. "The Importance of Being English: Jewish Immigration and the Decay of Liberal England." In *Metropolis: London: Histories and Representations Since 1800.* Edited by David Feldman and Gareth Stedman Jones, 56–84. London: Routledge, 1989.
Field, Geoffrey G. *Evangelist of Race: The Germanic Vision of Houston Stewart Chamberlain.* New York: Columbia University Press, 1981.
Fisch, Harold. *The Dual Image: The Figure of the Jew in English and American Literature.* New York: Ktav, 1971.
Fischer, Horst. *Judentum, Staat und Heer in Preussen im frühen 19. Jahrhundert: Zur Geschichte der Staatlichen Judenpolitik.* Tübingen: J. C. B. Mohr, 1968.
Fishberg, Maurice. "Die angebliche Rassen-Immunität der Juden." *ZDSJ* 4, no. 12 (1908): 177–188.
———. "Beiträge zur physischen Anthropologie der nordafrikanischen Juden." *ZDSJ* 1, no. 11 (1905): 1–5.
———. "The Comparative Pathology of the Jews." *New York Medical Journal* 73 (1901): 537–543, 576–582.
———. *The Jews: A Study of Race and Environment.* New York: Walter Scott Publishing, 1911.

———. "Materials for the Physical Anthropology of the Eastern European Jews." *Memoirs of the American Anthropological and Ethnological Societies* 1 (1905–1907).

———. "Physical Anthropology of the Jews: I: The Cephalic Index." *American Anthropologist*, n.s., 4 (1902): 684–706."

———. "Physical Anthropology of the Jews: II: Pigmentation." *American Anthropologist*, n.s., 5 (1903): 89–106.

———. "Probleme der Anthropologie der Juden: Zur Frage der Herkunft des blonden Elements im Judentum." *ZDSJ* 3, no. 1 (1907): 7–12; no. 2 (1907): 25–30.

———. "The Relative Infrequency of Tuberculosis among Jews." *American Medicine* 2 (1901): 695–699.

Fligier, C. "Zur Anthropologie der Semiten." *MAGW* 9 (1881): 247–253.

Flinders Petrie, W. M. "The Earliest Racial Portraits." *Nature: A Weekly Illustrated Journal of Science* 39 (1888): 128–130.

Forschungen zur Judenfrage. 8 vols. 1937–1943. (Arbeitstagung der Forschungsabteilung Judenfrage des Reichsinstituts für Geschichte des Neuen Deutschlands). Hamburg: Hanseatische Verlagsanstalt.

Franzos, Karl Emil. "Die Kolonisationsfrage." *AZJ* 55, no. 47 (1891).

Freimark, Peter. "Language Behaviour and Assimilation: The Situation of the Jews in Northern Germany in the First Half of the Nineteenth Century." *LBIYB* 24 (1979): 157–177.

Freud, Ernst L., ed. *The Letters of Sigmund Freud and Arnold Zweig.* New York: New York University Press, 1970.

Friedenthal, Hans. "Westasiaten und Europäer in anthropologischer Beziehung." *ZDSJ* 4, n.F., nos. 3–4, (1927): 40–45.

Fritz, George. *Die Ostjudenfrage: Zionismus und Grenzschluß.* Munich: Lehmann, 1915.

Galton, Francis. "Eugenics and the Jew." *Jewish Chronicle,* July 29, 1910.

———. *Hereditary Genius: An Inquiry into Its Laws and Consequences.* Cleveland: Meridian Books, 1962.

Gasman, Daniel. *The Scientific Origins of National Socialism: Social Darwinism in Ernst Haeckel and the German Monist League.* London: Macdonald, 1971.

Gay, Peter. *Freud, Jews and other Germans: Masters and Victims in Modernist Culture.* Oxford: Oxford University Press, 1979.

Gercke, Achim. *Die Rasse im Schriftum: Ein Wegweiser durch das rassenkundliche Schriftum.* Berlin: Alfred Metzner, 1933.

Gillispie, Charles Coulston, ed. *Dictionary of Scientific Biography.* Vol. 1. New York: Scribners, 1970.

Gilman, Sander L. *The Case of Sigmund Freud: Medicine and Identity at the Fin de Siècle.* Baltimore: Johns Hopkins University Press, 1993.

———. *Difference and Pathology: Stereotypes of Sexuality, Race, and Madness.* Ithaca: Cornell University Press, 1985.

———. *Freud, Race, and Gender*. Princeton, N.J.: Princeton University Press, 1993.

———. *Jewish Self-Hatred: Anti-Semitism and the Hidden Language of the Jews*. Baltimore: Johns Hopkins University Press, 1987.

———. *The Jews' Body*. New York: Routledge, 1991.

Giroux, Henry A. "Post-Colonial Ruptures and Democratic Possibilities." *Cultural Critique* 21 (Spring 1992): 5–39.

Glück, L. "Beiträge zur physischen Anthropologie der Spaniolen." *Wissenschaftliche Mitteilungen aus Bosnien und der Hercegovina* 4 (1896): 589–592.

Gobineau, Joseph Arthur Comte de. *The Inequality of Human Races*. New York: Howard Fertig, 1967.

Golb, Norman. "Notes on the Conversion of European Christians in the Eleventh Century." *Journal of Jewish Studies* 16 (1965): 69–74.

Golden, Peter B. *Khazar Studies: An Historico-Philological Inquiry into the Origins of the Khazars*. 2 vols. Budapest: Akademiai Kiado, 1980.

Goldstein, Eduard. "Des circonférences du thorax et de leur rapport a la taille." *Revue d'Anthropologie*, ser. 2, no. 7 (1885): 460–485.

Gould, Stephen Jay. *The Mismeasure of Man*. New York: Norton, 1981.

Graetz, Heinrich. *Die jüdischen Proselyten im Römerreiche unter den Kaisern Domitian, Nerva, Trajan und Hadrian*. Breslau: S. Schottlaender, 1884.

———. *The Structure of Jewish History and Other Essays*. Translated and edited by Ismar Schorsch. New York: Jewish Theological Seminary of America, 1975.

Greenberg, Louis. *The Jews in Russia*. 2 vols. New Haven: Yale University Press, 1951.

Greene, John C. *Science, Ideology, and World View: Essays in the History of Evolutionary Ideas*. Berkeley and Los Angeles: University of California Press, 1981.

Grégoire, Henri Baptiste. *An Essay on the Physical, Moral, and Political Reformation of the Jews*. London: C. Forster, Poultry, 1791.

Greive, Hermann. *Geschichte des modernen Antisemitismus in Deutschland*. Darmstadt: Wissenschaftliche Buchgesellschaft, 1983.

Grunwald, Max, ed. *Die Hygiene der Juden*. Dresden: N.p., 1911.

Günther, Hans F. K. *Rassenkunde des deutschen Volkes*. Munich: Lehmann, 1922.

———. *Rassenkunde des jüdischen Volkes*. Munich: Lehmann, 1930.

Gutmann, M.J. *Über den heutigen Stand der Rasse- und Krankheitsfrage der Juden*. Munich: Rudolph Müller & Steinicke, 1920.

Harris, Hugh. "The Fabulous Joseph Jacobs," *Jewish Chronicle*, no. 5 (1968).

Hartner, Herwig. *Erotik und Rasse: Eine Untersuchung über gesellschaftliche, sittliche und geschlechtliche Fragen mit Textillustrationen*. Munich: Deutscher Volksverlag, 1925.

Hertzberg, Arthur. *The French Enlightenment and the Jews*. New York: Columbia University Press, 1968.

Herz, Friedrich. *Race and Civilization*. N.p: Ktav, 1970.
Herzl, Theodor. *Briefe und Tagebücher*. Edited by Alex Bein, Hermann Greive, et al. Berlin: Propyläen Verlag, 1990.
———. *The Jewish State*. New York: Herzl Press, 1970.
Himmel [Heinrich?]. "Körpermessungen in der Bukowina." *MAGW* 18, n.F., (1888): 83–84.
Hofacker, Johann Daniel. *Über die Eigenschaften welche sich bei Menschen und Thieren von den Eltern auf die Nachkommen vererben*. Tübingen: Osiander, 1828.
Holmes, Colin. *Anti-Semitism in British Society, 1876–1939*. New York: Holmes and Meier, 1979.
Hoppe, H. *Krankheiten und Sterblichkeit bei Juden und Nichtjuden: Mit besonderer Berücksichtigung der Alkoholfrage*. Berlin: S. Calvary, 1903.
Hough, John S. "Longevity and Biostatic Peculiarities of the Jewish Race." *Medical Record* (1873): 241–244.
Huth, A.H. *The Marriage of Near Kin Considered with Respect to the Laws of Nations*. London: J & A. Churchill, 1875.
Huxley, Henry Minor. "The Samaritans" *Jewish Encyclopedia* 10:674–676.
———. "Zur Anthropologie der Samaritaner." *ZDSJ* 2, nos. 8 and 9 (1906): 137–139.
Hyamson, Albert M. "The Lost Tribes and the Return of the Jews to England" *TJHSE* 5 (1902-1905): 115–147.
Hygiene und Judentum: Eine Sammelschrift. Dresden: JAC Sternlicht, 1930.
Ikow, Constantine. "Neue Beiträge zur Anthropologie der Juden." *AA* 15 (1884): 369–389.
Jacobs, Joseph. *Jewish Ideals and Other Essays*. New York: Macmillan, 1896.
———. *The Jewish Question, 1877–1884; Bibliographical Hand-List*. London: Trübner, 1885.
———. *The Jewish Race: A Study in National Character*. London: Privately printed, 1899.
———. *Jewish Statistics: Social, Vital and Anthropometric*. London: D. Nutt, 1891.
———. *Jews of Distinction*. N.p., 1919.
———. *The Persecution of the Jews of Russia*. London: Wertheimer, Lea, 1890.
———. "Are Jews Jews?" *Popular Science Monthly* 55 (1899): 502–511.
———. *An Inquiry into the Sources of the History of the Jews in Spain*. London: Nutt, 1894.
———. *Jewish Contributions to Civilization: An Estimate*. Philadelphia: Jewish Publication Society, 1945.
———. "Die Juden in den Vereinigten Staaten." *ZDSJ* 2, no. 3 (1906): 33–39; no. 4 (1906): 54–57.
———. "Jüdische Volkskunde und die Einleitung der Statistik der Juden." In *Jüdische Statistik: Eine Sammelschrift*, 30–35. Berlin: Jüdischer Verlag, 1918.

———. "Little St. Hugh of Lincoln: Researches in History, Archaeology, and Legend." *TJHSE* 1 (1893–1894): 89–135.
Jacobs, Joseph, and Hermann Landau. *Yiddish-English Manual*. London: E. W. Rabbinowicz, 1893.
Jacobs, Joseph, and Lucien Wolf. *Bibliotheca Anglo-Judaica*. London: William Clowes, 1887.
Jeiteles, Israel. *Die Kultusgemeinde der Israeliten in Wien*. Vienna: L. Rosner, 1873.
Jellinek, Adolf. *Der Jüdische Stamm: Ethnographische Studien*. Vienna: Herzfeld & Bauer, 1869.
Jewish Encyclopaedia. 12 vols. New York: Funk & Wagnalls, 1901–1906.
A Jewish Scholar's Career: The Maccabaeans: A Report of the Speeches at the Dinner to Mr. Joseph Jacobs. London: Jewish Chronicle Office, 1896.
"Der Judenstamm in naturhistorischer Betrachtung." *Das Ausland* 53 (1880): no. 23, 453–456; no. 24, 474–476; no. 25, 483–488; no. 26, 509–513; no. 27, 536–539.
Jüdische Bevölkerungs-Politik: Bericht über die Tagung des bevölkerungspolitischen Ausschusses des preussischen Landesverbandes jüdischer Gemeinden vom 24. Februar 1929—Material zur jüdischen Bevölkerungspolitik. Berlin: N.p., 1929.
Jüdisches Lexikon. 5 vols. Berlin: Jüdischer Verlag, 1927–1930.
Judt, I. M. *Die Juden als Rasse*. Berlin: Jüdischer Verlag, 1903.
Jung, Leo, ed. *Men of the Spirit*. New York: Kymson Publishing, 1964.
Jungmann, Max. "Ist das jüdische Volk degeneriert?" *Die Welt* (1902): 3–4.
Kahn, Fritz. *Die Juden als Rasse und Kulturvolk*. Berlin: Welt-Verlag, 1921.
Kampe, Norbert. "Jews and Anti-Semites at Universities in Imperial Germany (I)—Jewish Students: Social History and Social Conflict." *LBIYB* 30 (1985): 357–394.
———. "Jews and Anti-Semites at Universities in Imperial Germany (II)—The Friedrich-Wilhelms-Universität of Berlin: A Case Study on the Students' 'Jewish Question.'" *LBIYB* 32 (1987): 43–101.
Kampmann, Wanda. *Deutsche und Juden: Die Geschichte der Juden in Deutschland vom Mittelalter bis zum Beginn des Ersten Weltkrieg*. Frankfurt am Main: Fischer, 1986.
Katz, Albert. *Die Juden im Kaukasus*. Berlin: Hugo Schildberger, 1894.
Katz, David S. *Philo-Semitism and the Readmission of the Jews to England, 1603–1635*. Oxford: Clarendon Press, 1982.
Katz, Jacob. *The Darker Side of Genius: Richard Wagner's Anti-Semitism*. Hanover, N.H.: University of New England Press, 1986.
———. *From Prejudice to Destruction: Anti-Semitism, 1700–1933*. Cambridge, Mass.: Harvard University Press, 1980.
———. *Jewish Emancipation and Self-Emancipation*. Philadelphia: Jewish Publication Society, 1986.

———. *Out of the Ghetto: The Social Background of Jewish Emancipation, 1770–1870.* Cambridge, Mass.: Harvard University Press, 1973.

Kaufmann, David. "George Eliot und das Judentum." In *Gesammelte Schriften,* 1:39–79. Frankfurt: M. Brann, 1908.

Kautsky, Karl. *Are the Jews a Race?* New York: International Publishers, 1926.

Kaznelson, Paul. "Über einige 'Rassenmerkmale' des jüdischen Volkes." *ARGB* 10, no. 4 (1913): 484–502.

Kerr, Norman. *Inebriety: Its Etiology, Pathology, Treatment and Jurisprudence.* London: Lewis, 1888.

Kevles, Daniel. *In the Name of Eugenics: Genetics and the Uses of Human Heredity.* New York: Knopf, 1985.

Klausner, Yisrael. *Opositsia le-Herzl.* Jerusalem: Archiever, 1960.

Klein, Dennis B. *Jewish Origins of the Psycho-Analytic Movement.* Chicago: University of Chicago Press, 1985.

Klemm, Gustav. *Allgemeine Cultur-Geschichte der Menschheit.* Vol. 1. Leipzig: B. G. Teubner, 1843.

Knox, Robert. *Races of Men: A Fragment.* Miami: Mnemosyne, 1969.

Kobler, Franz. *Juden und Judentum in deutschen Briefen aus drei Jahrhunderten.* Königstein/Ts: Jüdischer Verlag Athenäum, 1984.

Kollmann, Julius. "Die Rassenanatomie der Hand und die Persistenz der Rassenmerkmale." *AA* 28 (1903): 91–141.

———. "Schädel und Skeletreste aus einem Judenfriedhof des 13. und 14. Jahrhunderts zu Basel." *Verhandlungen der naturforschenden Gesellschaft zu Basel* 7 (1885): 648–656.

———. "Zur Anthropologie der Juden." *Correspondenzblatt der deutschen Gesellschaft für Anthropologie, Ethnologie und Urgeschichte* 48 (1917): 1–5.

Koralik, J. "Untergang des Judentums? Eine Erwiderung." *OSE-Rundschau: Zeitschrift der Gesellschaft für Gesundheitsschutz der Juden* 1 (1929): 1–6; and 2 (1929): 1–8.

Kraepelin, Emil. *Psychiatrie: Ein Lehrbuch.* Vol. 1. Leipzig: Johann Ambrosius Barth, 1903.

———. "Zur Entartungsfrage." *Zentralblatt für Nervenheilkunde und Psychiatrie* 31 (October 1908): 745–749.

Krafft-Ebing, Richard von. *Text-Book of Insanity.* Philadelphia: F. A. Davis, 1904.

Kümmel, Werner Friedrich. "Die Ausschaltung." *Deutsches Ärzteblatt—Ärztliche Mitteilungen* 85, no. 33 (1988): 1–4.

———. "Jüdische Ärzte in Deutschland zwischen Emanzipation und 'Ausschaltung.'" In *Richard Koch und die ärztliche Diagnose.* Edited by Gert Preiser, 15–47. Hildesheim: Olms, 1988.

———. "Virchow und der Antisemitismus." *Medizinhistorisches Journal* 3 (1968): 165–179.

Lagneau, Gustave. "Sur la race juive et sa pathologie." *Bulletin de la Société d'anthropologie de Paris* 2 (1891): 539–549, 556–557.

Lamberti, Marjorie. "From Coexistence to Conflict: Zionism and the Jewish Community in Germany, 1897–1914." *LBIYB* 27 (1982): 53–85.

Langbehn, Julius. *Rembrandt als Erzieher.* Leipzig: C. L. Hirschfeld, 1909.

Lask Abrahams, Beth-Zion. "George Eliot: Her Jewish Associations—A Centenary Tribute," *TJHSE* 26 (1974–1978): 53–61.

Latham, Robert Gordon. *The Natural History of the Varieties of Man.* London: John van Voorst, 1850.

Lawrence, William. *Lectures on Physiology, Zoology, and the Natural History of Man.* London: James Smith, 1823.

Le Goyt, Alfred. *De certaines immunités biostatiques de la race juive.* Paris: Bureau des archives israelites, 1868.

Leschnitzer, Adolf. *The Magic Background of Modern Anti-Semitism: An Analysis of the German-Jewish Relationship.* New York: International Universities Press, 1956.

Lesky, Erna. *The Vienna Medical School in the Nineteenth Century.* Baltimore: Johns Hopkins University Press, 1976.

Lestschinsky, Jacob. *Dos yidishe folk in tsifern.* Berlin: Klal Verlag, 1922.

———. *Das wirtschaftliche Schicksal des deutschen Judentums: Aufstieg, Wandlung, Krise, Ausblick.* Berlin: Energiadruck, 1932.

Lestschinsky, Jacob; B. Brutzkus; and Jacob Segall, eds. *Bleter far yidishe demographie, statistik un ekonomik.* Berlin: N.p., 1925.

Levy, Richard S. *The Downfall of the Anti-Semitic Political Parties in Imperial Germany.* New Haven: Yale University Press, 1975.

Lewy, Joseph. "Antisemitismus und Medizin." *Im deutschen Reich* 5, no. 1 (1899): 1–19.

Lifschutz, Ezekiel. "Jacob Gordin's Proposal to Establish an Agricultural Colony." *American Jewish Historical Quarterly* 56 (1966–1967): 151–162.

Lipman, V. D. *Jewish Chronicle,* February 5, 1982.

———. *Social History of the Jews in England, 1850–1950.* London: Watts, 1954.

Loewe, Heinrich. *Proselyten: Ein Beitrag zur Geschichte der jüdischen Rasse.* Berlin: Soncino-Gesellschaft, 1926.

Lombroso, Cesare. *Der Antisemitismus und die Juden im Lichte der modernen Wissenschaft.* Leipzig: Georg H. Wigand, 1894.

Lonsdale, Henry. *A Sketch of the Life and Writings of Robert Knox, the Anatomist.* London: Macmillan, 1870.

Lovejoy, Arthur O. *The Great Chain of Being: A Study of the History of an Idea.* New York: Harper Torchbooks, 1960.

Low, D. *Jews in the Eyes of Germans: From the Enlightenment to Imperial Germany.* Philadelphia: Institute for the Study of Human Issues, 1979.

Lowenstein, Steven M. *The Mechanics of Social Change: Essays in the Social History of German Jewry.* Atlanta: Scholars Press, 1992.

———. "The Yiddish Written Word in Nineteenth-Century Germany." *LBIYB* 24 (1979): 179–192.

Mann, Gunter. "Biologie und Geschichte: Ansätze und Versuche zur biologistischen Theorie der Geschichte im 19. und beginnenden 20. Jahrhundert." *Medizinhistorisches Journal* 10, no. 4 (1975): 281–306.

———. "Dekadenz-Degeneration-Untergangsangst im Lichte der Biologie des 19. Jahrhunderts." *Medizinhistorisches Journal* 20, nos. 1/2 (1985): 6–35.

Marcus, Alfred. *Die wirtschaftliche Krise des deutschen Juden: Eine soziologische Untersuchung.* Berlin: Georg Stilke, 1931.

Marmor, Kalmon. *Jacob Gordin.* New York: Yidisher Kultur Farband, 1953.

Marrus, Michael. *The Politics of Assimilation: The French Jewish Community at the Time of the Dreyfus Affair.* Oxford: Oxford University Press, 1971.

Marx-Engels Werke. Vol. 30. Berlin: Dietz Verlag, 1964.

Massing, Paul W. *Rehearsal for Destruction: A Study of Political Anti-Semitism in Imperial Germany.* New York: Howard Fertig, 1967.

Maurer, Friedrich. "Mitteilungen aus Bosnien." *Das Ausland* 49 (1869): 1161–1164; and 50 (1869): 1183–1185.

Mayr, Georg. "Die bayerische Jugend nach der Farbe der Augen, der Haare und der Haut." *Zeitschrift des königlich bayerischen statistischen Bureaus* 7 (1875): 273–311.

Mendes-Flohr, Paul R., and Judah Reinharz, eds. *The Jew in the Modern World: A Documentary History.* New York: Oxford University Press, 1980.

Messer, Ellen. "Franz Boas and Kaufmann Kohler: Anthropology and Reform Judaism." *Jewish Social Studies* 48, no. 2 (1986): 127–140.

Meyer, Eduard. *Die Entstehung des Judenthums.* Halle: Max Niemeyer, 1896.

———. *Die Israeliten und ihre Nachbarstämme.* Halle: Max Niemeyer, 1906.

Meyer, Michael. *The Origins of the Modern Jew: Jewish Identity and European Culture in Germany, 1749–1824.* Detroit: Wayne State University Press, 1967.

———. *Response to Modernity: A History of the Reform Movement in Judaism.* New York: Oxford University Press, 1988.

Michaelis, Curt. "Die jüdische Auserwählungsidee und ihre biologische Bedeutung." *ZDSJ* 1, no. 2 (1905): 1–4.

Morantz-Sanchez, Regina Markell. *Sympathy and Science: Women Physicians in American Medicine.* New York. Oxford University Press, 1987.

Mosse, George L. *The Crisis of German Ideology: Intellectual Origins of the Third Reich.* New York: Grosset & Dunlap, 1964.

———. *German Jews beyond Judaism.* Bloomington: Indiana University Press, 1985.

———. *Germans and Jews: The Right, the Left and the Search for a Third Force in Pre-Nazi Germany.* New York: Howard Fertig, 1970.

———. *Masses and Man: Nationalist and Fascist Perceptions of Reality.* Detroit: Wayne State University Press, 1987.

———. *Nationalism and Sexuality: Respectability and Abnormal Sexuality in Modern Europe.* New York: Howard Fertig, 1985.

———. *Toward the Final Solution: A History of European Racism.* Madison: University of Wisconsin Press, 1985.

Mosse, Werner, and Arnold Paucker, eds. *Deutsches Judentum in Krieg und Revolution, 1916–1923.* Tübingen: J. C. B. Mohr, 1971.

———. *Entscheidungsjahr 1932: Zur Judenfrage in der Endphase der Weimarer Republik.* Tübingen: J. C. B. Mohr, 1966.

———. *Juden in wilhelminischen Deutschland 1890–1914.* Tübingen: J. C. B. Mohr, 1976.

Mühlmann, Wilhelm E. *Geschichte der Anthropologie.* Wiesbaden: Aula, 1986.

Nagel, E. "Der hohe Knabenüberschuss der Neugeborenen der Jüdinnen." *Statistische Monatschrift* 10 (1884): 183–186.

Nagl, Erasmus. *Die nachdavidische Königsgeschichte Israels: Ethnographisch und geographisch beleuchtet.* Vienna: Carl Fromme, 1905.

Neubauer, Adolf. "Notes on the Race-Types of the Jews." *Journal of the Royal Anthropological Institute of Great Britain and Ireland* 15 (1885): 17–23.

Nichols, J. B. "The Numerical Proportion of the Sexes at Birth." *Memoirs American Anthropological Association* 1 (1907): 249–300.

Niewyk, Donald L. *The Jews in Weimar Germany.* Baton Rouge, 1980.

Nöldeke, Theodore. *Sketches from Eastern History.* Beirut: Khayats, 1963.

Nordau, Max. *The Conventional Lies of Our Civilization.* New York: Arno Press, 1975.

———. *Degeneration.* New York: Howard Fertig, 1968.

———. *Zionistische Schriften.* Berlin: Jüdischer Verlag, 1923.

Nossig, Alfred. "Die Auserwähltheit der Juden im Lichte der Biologie." *ZDSJ* 1, no. 3 (1905): 1–5.

Nossig, Alfred, ed. *Jüdische Statistik.* Berlin: Jüdischer Verlag, 1903.

Nott, J. C., and Gliddon, G. R. *Types of Mankind; or, Ethnological Researches.* Philadelphia: Lippincott, Grambo, 1854.

Noy, Dov. "Dr. Max Grunwald—The Founder of Jewish Folkloristics." *Folklore Research Center Studies* 4 (1982): i–ix.

———. "Eighty Years of Jewish Folkloristics: Achievements and Tasks." In *Studies in Jewish Folklore.* Edited by Frank Talmage, 1–11. Cambridge, Mass.: Association for Jewish Studies, 1980.

Nussbaum, William. "Anthropological Studies on German Jews (1933/34)." Mimeograph. New York: Leo Baeck Institute.

Nyström, Anton. *Über Formenveränderungen des menschlichen Schädels und deren Ursachen."* AA 27 (1902): 211–231.

Oppenheimer, Franz. "Stammesbewusstsein und Volksbewusstsein." *Die Welt* 14, no. 7 (1910): 139–143.

Oxaal, Ivaar, Michael Pollack, and Gerhard Botz, eds. *Jews, Antisemitism and Culture in Vienna.* London: Routledge & Kegan Paul, 1987.

Oxaal, Ivaar, and Walter R. Weitzmann. "The Jews of Pre-1914 Vienna: An Exploration of Basic Sociological Dimensions." *LBIYB* 30 (1985): 395–432.

References

Panitz, Esther L. *The Alien in their Midst: Images of Jews in English Literature.* Rutherford, N.J.: Fairleigh Dickinson University Press, 1981.

Pasmanik, Daniel. "Die Judenassimilation seit Mendelssohn." *Jüdischer Almanach 5670* [Bar-Kochba in Wien], 50–65. Vienna: Jüdischer Verlag, 1910.

——. *Die Seele Israels: Zur Psychologie des Diasporajudentums.* Cologne: Jüdischer Verlag, 1911.

Patai, Raphael. *The Jewish Mind.* New York: Scribners, 1977.

Patai, Raphael, and Jennifer P. Wing. *The Myth of the Jewish Race.* New York: Scribners, 1975.

Pearson, Karl. *The Life, Letters and Labours of Francis Galton.* Vol. 2. Cambridge: Cambridge University Press, 1924.

Perier, J. A. N. "Essai sur les croisements ethniques." *Mémoires de la Société d'Anthropologie de Paris* 1–2 (1860–1865). Part 1, 69–92; and part 2, 187–236.

Perspectives of German-Jewish History in the Nineteenth and Twentieth Century. [Proceedings of a conference held at the Leo Baeck Institute, Jerusalem, summer 1970.] Jerusalem: Jerusalem Academic Press, 1971.

Phillips, Olga Somech. "Joseph Jacobs Centenary." *Jewish Chronicle,* no. 4 (1954).

Picciotto, James. *Sketches of Anglo-Jewish History.* London: Soncino Press, 1956.

Platter, J. "Die Hofacker-Sadler'sche Hypothese im Lichte der österreichischen Bevölkerungs-Statistik." *Statistische Monatschrift* 1 (1875): 451–456.

Ploetz, Alfred. *Die Tüchtigkeit unserer Rasse und der Schutz der Schwachen.* Berlin: S. Fischer, 1895.

Poliakov, Leon. *The Aryan Myth: A History of Racist and Nationalist Ideas in Europe.* New York: Meridian, 1977.

——. *The History of Anti-Semitism.* 4 vols. New York: Vanguard Press, 1965–1975.

Pollins, Harold. *Economic History of the Jews in England.* Rutherford, N.J: Fairleigh Dickinson University Press, 1982.

Poole, W.H. *Anglo-Israel; or, The Saxon Race Proved to be the Lost Tribes of Israel.* London: Robert Banks, 1889.

Poppel, Stephen M. *Zionism in Germany, 1897–1933: The Shaping of Jewish Identity.* Philadelphia: Jewish Publication Society, 1977.

Preston, David Lawrence. "Science, Society, and the German Jews: 1870–1933." Ph.D. diss., University of Illinois, Urbana-Champaign, 1971.

Prichard, James Cowles. *Researches into the Physical History of Man.* Edited by George W. Stocking, Jr. Chicago: University of Chicago Press, 1973.

Proctor, Robert N. "From *Anthropologie* to *Rassenkunde* in the German Anthropological Tradition." In *Bones, Bodies, Behavior: Essays on Biological Anthropology.* Edited by George W. Stocking, Jr., 138–179. Madison: University of Wisconsin Press, 1988.

———. *Racial Hygiene: Medicine under the Nazis*. Cambridge, Mass.: Harvard University Press, 1988.
Programm und Organisations-Statut der Demokratischen Zionistischen Fraktion (1902).
Pulzer, Peter G. J. *The Rise of Political Anti-Semitism in Germany and Austria*. New York: John Wiley, 1964.
Querner, H. "Zur Geschichte der Anthropologie." *Anthropologischer Anzeiger* 44, no. 12 (1986): 281–296.
The Race Question in Modern Science. New York: UNESCO, Whiteside and William Morrow, 1956.
Raciborski, Adam. *Traité de la Mentruation*. Paris: J. B. Baillière, 1868.
Rae, Isobel. *Knox, the Anatomist*. Edinburgh and London: Oliver and Boyd, 1964.
Ranke, J. "Zur Statistik und Physiologie der Körpergrösse der bayerischen Militärpflichtigen." *Beiträge zur Anthropologie Bayerns* 4 (1881): 1–35.
Reibmayr. Albert. *Inzucht und Vermischung beim Menschen*. Leipzig: Franz Deuticke, 1897.
Reinharz, Jehuda. *Fatherland or Promised Land: The Dilemma of the German Jew, 1893–1914*. Ann Arbor: University of Michigan Press, 1975.
Reisin, Zalman. *Leksikon fun der nayer yidisher literatur*. New York: Congress for Jewish Culture, 1960.
Remondino, P.C. *History of Circumcision*. Philadelphia and New York: F. A. Davis, 1900.
Renan, Ernest. *History of the People of Israel*. Vol. 5. Boston: Roberts Brothers, 1895.
———. *Le judaisme comme race et comme religion*. Paris: Michel Levy Frères, 1883.
"Report of a Lecture by B. W. Richardson, 'The Mosaic Sanitary Code and Its Effect on the Jewish Race, 25th March, 1876.'" Mocatta boxed pamphlet A 99 RIC., Mocatta Library, University College, London.
Report of House of Lords Select Committee on Sweating System (1888–89).
Report of Select Committee of House of Commons on Immigration and Emigration (1888).
Ripley, William Z. *The Races of Europe*. New York: D. Appelton, 1899.
———. "The Racial Geography of Europe: A Sociological Study. Supplement: The Jews." *Popular Science Monthly* 54 (January, 1899). Part 1: 163–175; part 2: 338–351.
"Ripley über die Anthropologie der Juden." *Globus* 76, no. 2 (1899): 21–27.
Ritterband, Paul, ed. *Modern Jewish Fertility*. Leiden: E. J. Brill, 1981.
Rosenbloom, Joseph R. *Conversion to Judaism: From the Biblical Period to the Present*. Cincinnati: Hebrew Union College Press, 1978.
Rosenfeld, Siegfried. "Todesursachen bei Juden in österreichischen Städten." *ZDSJ* 3, no. 11 (1907): 161–167.

Rossiter, Margaret. *Women Scientists in America: Struggles and Strategies to 1940.* Baltimore: Johns Hopkins University Press, 1982.
Rost, L. *Zur Berufsthätigkeit der Juden: Gegen den Vorwurf ihrer Arbeitsscheu.* Alzey: A. Meschett, 1880.
Rotenstreich, Nathan. *The Recurrent Pattern: Studies in Anti-Judaism in Modern Thought.* London: Weidenfeld & Nicolson, 1963.
Roth, Cecil. *A History of the Jews in England.* Oxford: Clarendon Press, 1949.
———. *A Short History of the Jewish People.* London: East and West Library, 1948.
Rozenblit, Marsha L. "The Assertion of Identity. Jewish Student Nationalism at the University of Vienna before the First World War." *LBIYB* 27 (1982): 171–186.
———. *The Jews of Vienna, 1867–1914: Assimilation and Identity.* Albany: State University of New York Press, 1983.
Rudolphi, D. Karl Asmund. *Beyträge zur Anthropologie und allgemeinen Naturgeschichte.* Berlin: Haude and Spener, 1812.
Ruppin, Arthur. *The Jewish Fate and Future.* London: Macmillan, 1940.
———. *The Jews in the Modern World.* London: Macmillan, 1934.
———. *The Jews of Today.* New York: Henry Holt, 1913.
———. *Die Juden der Gegenwart: Eine sozialwissenschaftliche Studie.* Cologne: Jüdischer Verlag, 1911.
———. "Die jüdische Rassenfrage." *ZDSJ* 3, nos. 8 and 9 (1907): 138–140.
———. *Memoirs, Diaries, Letters.* Edited by Alex Bein. London: Weidenfeld & Nicolson, 1971.
———. "Die Mischehe." *ZDSJ* 4, no. 2 (1908): 17–23.
———. "Der Rassenstolz der Juden." *ZDSJ* 6, no. 6 (1910): 88–92.
———. *Soziologie der Juden.* 2 vols. Berlin: Jüdischer Verlag, 1930–31.
———. "Der Stand der Statistik der Juden." *ZDSJ* 3, no. 12 (1907): 177–78.
———. *Tagebücher, Briefe, Erinnerungen.* Jüdischer Verlag Athenäum: Königstein/Ts., 1985.
Rutland, Suzanne D. *Edge of the Diaspora: Two Centuries of Jewish Settlement in Australia.* Sydney: Collins, 1988.
Salbstein, M. C. N. *The Emancipation of the Jews in Britain: The Question of the Admission of the Jews to Parliament, 1828–1860.* Rutherford, N.J.: Farleigh Dickinson University Press, 1982.
Salomon, Julius. "Eheschliessungen zwischen Juden und Christen in Kopenhagen in den Jahren 1880–1903." *ZDSJ* 1, no. 1 (1905): 5–6.
Samter, Nathan. *Judentaufen im 19. Jahrhundert.* Berlin: M. Poppelauer, 1906.
———. *Judenthum und Proselytismus.* Breslau: W. Jacobsohn, 1897.
Samuelson, James. *A History of Strong Drink: A Review Social, Scientific and Political.* London: Trübner, 1878.
Sandler, Aron. *Anthropologie und Zionismus: Ein populär wissenschaftlicher Vortrag.* Breslau: Jüdischer Buch- und Kunstverlag, 1904.

———. "Mischehe und jüdisch-nationale Gesinnung." *Jüdische Rundschau*. April 22 (1904).

———. "Noch einmal die Mischehe." *Jüdische Rundschau*, May 20, 1904.

Sayce, A. H. *The Races of the Old Testament*. London: Religious Tract Society, 1891.

Scheiber, S. H. "Untersuchungen über den mittleren Wuchs der Menschen in Ungarn." *AA* 13 (1881): 133–267.

Scheuer, Oscar Franz. *Burschenschaft und Judenfrage: Der Rassenantisemitismus in der deutschen Studentenschaft*. Berlin: Verlag Berlin-Wien, 1927.

———. *Die geschichtliche Entwicklung des deutschen Studententums in Österreich mit besonderer Berücksichtigung der Universität Wien von ihrer Gruendung bis zur Gegenwart*. Vienna: Ed. Beyers, 1910.

Schimmer, G.A. "Erhebungen über die Farbe der Augen, der Haare und der Haut bei den Schulkindern Österreiches." *MAGW*, suppl. 1 (1884).

Schlesinger, Bella, ed. *Führer durch die Jüdische Wohlfahrtspflege in Deutschland*. Berlin: Fritz Scherbel, 1928.

Schnitzler, Arthur. *Professor Bernhardi: A Comedy in Five Acts*. New York: Simon and Schuster, 1928.

Scholem, Gershom. *On Jews and Judaism in Crisis: Selected Essays*. New York: Schocken Books, 1976.

Schorsch, Ismar. "From Wolfenbüttel to Wissenschaft: The Divergent Paths of Isaak Markus Jost and Leopold Zunz." *LBIYB* 22 (1977): 109–128.

———. *Jewish Reactions to German Anti-Semitism, 1870–1914*. New York: Columbia University Press, 1972.

———. "The Myth of Sephardic Supremacy." *LBIYB* 34 (1989): 47–66.

Schorske, Carl E. *Fin-de-Siécle Vienna: Politics and Culture*. New York: Vintage Books, 1979.

Schüler, Alexander. *Der Rassenadel der Juden: Der Schlüssel zur Judenfrage*. Berlin: Jüdischer Verlag, 1912.

Searle, Geoffrey R. *Eugenics and Politics in Britain, 1900–1914*. Leyden: Noordhoff, 1976.

———. *The Quest for National Fitness: A Study in British Politics and Political Thought, 1899–1914*. Oxford: Basil Blackwell, 1971.

Seltzer, Robert M. "Joining the Jewish People from Biblical to Modern Times." In *Pushing the Faith: Proselytism and Civility in a Pluralistic World*. Edited by Martin E. Marty and Frederick E. Greenspahn, 41–63. New York: Crossroad, 1988.

Sharot, Stephen. "Hasidism in Modern Society." In *Essential Papers on Hasidism*. Edited by Gershon Hundert, 511–531. New York: New York University Press, 1991.

Siebert, Friedrich. *Der völkische Gedanke und die Verwirklichung des Zionismus: Eine Betrachtung zur Versöhnung und zur Scheidung der Völker*. Munich: Lehmann, 1916.

Singer, Heinrich. *Allgemeine und spezielle Krankheitslehre der Juden*. Leipzig: Benno Konegen, 1904.

Smith, Samuel Stanhope. *An Essay on the Causes of the Variety of Complexion and Figure in the Human Species*. Cambridge, Mass.: Belknap Press, Harvard University Press, 1965.

Snyder, Louis. *Race: A History of Modern Ethnic Theories*. New York: Longmans, Green, 1939.

Sofer, Leo. "Armenier und Juden." *ZDSJ* 3, no. 5 (1907): 65–69.

———. "Über die Entmischung der Rassen." *ZDSJ* 1, no. 10 (1905): 9–12.

———. "Welcher Rasse gehörte Jesus an?" *ZDSJ* 5, no. 6 (1909): 81–87.

———. "Zur Biologie und Pathologie der jüdischen Rasse." *ZDSJ* 2, no. 6 (1906): 85–92.

Sombart, Werner. *The Jews and Modern Capitalism*. Glencoe, Ill.: Free Press, 1951.

Sorkin, David. *The Transformation of German Jewry, 1780–1840*. New York: Oxford University Press, 1987.

Spector, Mordechai, ed. *Der Hoyzfraynd*. Warsaw: N.p., 1889.

Stepan, Nancy. *The Idea of Race in Science: Great Britain, 1800–1960*. London: Macmillan, 1982.

Sterling, Eleonore. *Judenhaß: Die Anfänge des politischen Antisemitismus in Deutschland 1815–1850*. Frankfurt am Main: Europäische Verlagsanstalt.

Stern, Fritz. *Gold and Iron: Bismarck, Bleichröder and the Building of the German Empire*. London: George Allen & Unwin, 1977.

Stieda, Ludwig. "Ein Beitrag zur Anthropologie der Juden." *AA* 14 (1883): 61–71.

Stocking, George W., Jr. "What's in a Name? The Origins of the Royal Anthropological Institute (1837–71)." *Man: The Journal of the Royal Anthropological Institute* 6, no. 3 (1971): 369–390.

———. *Victorian Anthropology*. New York: Free Press, 1987.

———. *Race, Culture, and Evolution: Essays in the History of Anthropology*. New York: Free Press, 1968.

Stocking, George W., Jr., ed. *Bones, Bodies, Behavior: Essays on Biological Anthropology*. Madison: University of Wisconsin Press, 1988.

Stoddard, Lothrop. *Racial Realities in Europe*. New York: Scribners, 1924.

Stone, Harry. "From Fagin to Riah: Jews and the Victorian Novel." *Midstream* 6, no. 1 (1960): 21–37.

Stratz, C. H. *Was sind Juden? Eine ethnographisch-anthropologische Studie*. Vienna: F. Tempsky, 1903.

Tal, Uriel. *Christians and Jews in Germany: Religion, Politics, and Ideology in the Second Reich, 1870–1914*. Ithaca: Cornell University Press, 1975.

Theilhaber, Felix A. "Beiträge zur jüdischen Rassenfrage." *ZDSJ* 6, no. 3 (1910): 40–44.

———. *Die Beschneidung*. Berlin: Louis Lamm, 1927.

———. "Bevölkerungsproblematische Phantasien und Bevölkerungstheoretische

Illusionspolitik." *OSE-Rundschau* 3, no. 11 (1928): 1–7.

———. *Dein Reich komme! Ein chiliastischer Roman aus der Zeit Rembrandts und Spinozas.* Berlin: C. A. Schwetschke, 1924.

———. "Gerim in Palästina." *Die Welt,* no. 27 (1907).

———. *Geschichte des jüdischen Volkes.* 2 vols. Berlin: Kedem-Verlag, 1936.

———. *Goethe. Sexus und Eros.* Berlin-Grunewald: Horen-Verlag, 1929.

———. *Historischer Atlas Israels in Erez Israel und der Diaspora.* Tel Aviv: Dr. J. Szapiro, 1946.

———. *Die Juden im Weltkrieg; mit besonderer Berücksichtigung der Verhältnisse für Deutschland.* Berlin: Weltverlag, 1916.

———. *Judenschicksal.* Tel Aviv: Olympia, 1946.

———. *Jüdische Flieger im Kriege, ein Blatt der Erinnerung.* Berlin: Louis Lamm, 1919.

———. *Die Schädigung der Rasse durch soziales und wirtschaftliches Aufsteigen, bewiesen an den Berliner Juden.* Berlin: Louis Lamm, 1914.

———. *Schicksal und Leistung. Juden in der deutschen Forschung und Technik.* Berlin: Welt-Verlag, 1931.

———. *Schlichte Kriegserlebnisse.* Berlin: Louis Lamm, 1916.

———. "Die Sterblichkeit des jüdischen Nachwuchses und die Geburtenfrage." In *Hygiene und Judentum: Eine Sammelschrift.* Dresden: N.p., 1930.

———. *Das sterile Berlin, eine volkswirtschaftliche Studie.* Berlin: Eugen Marquardt, 1913.

———. "Der Untergang der deutschen Juden." *Jüdisches Literature-Blatt* 33 (1911): 185–188.

———. *Der Untergang der deutschen Juden: Eine volkswirtschaftliche Studie.* Berlin: Jüdischer Verlag, 1921.

———. "Zur Lehre von dem Zusammenhang der sozialen Stellung und der Rasse mit der Entstehung der Uteruscarcinome." Inaugural dissertation. Munich: K. K. Ludwigs-Maximilians Universität, 1910.

Thon, Jacob. "Kriminalität der Christen und Juden in Oesterreich." *ZDSJ* 2, no. 1 (1906): 6–10.

Toland, John. *Reasons for Naturalizing the Jews in Great Britain and Ireland.* Jerusalem: Hebrew University, 1963.

Toury, Jacob. *Soziale und politische Geschichte der Juden.* Dusseldorf: Droste, 1977.

Trachtenberg, Joshua. *The Devil and the Jews: The Medieval Conception of the Jew and Its Relation to Modern Antisemitism.* Cleveland: Meridian Books, 1961.

Virchow, Rudolf. "Gesamtbericht über die von der deutschen anthropologischen Gesellschaft veranlassten Erhebungen über die Farbe der Haut, der Haare und der Augen der Schulkinder in Deutschland." *AA* 16 (1886): 275–475.

———. "Die Mumien der Könige im Museum zu Balacq." *Sitzungs-Berichte der Berlin Akademie der Wissenschaft* (1888): 767–787.

———. "Über Erblichkeit I: Die Theorie Darwin's." *Deutsche Jahrbücher für Politik und Literatur* 6 (1863): 339–358.
Vogt, Karl. *Lectures on Man: His Place in Creation, and in the History of the Earth.* London: Longman, Green, Longman, & Roberts, 1864.
Voisin, A. "Contribution a l'histoire des mariages entre consanguins." *Mémoires de la Société d'Anthropologie de Paris* 2 (1865): 433–459.
Volkov, Shulamit. *Jüdisches Leben und Antisemitismus im 19. und 20. Jahrhundert.* Munich: C. H. Beck, 1990.
Von Erckert, R. *Der Kaukasus und seine Völker.* Leipzig: Eduard Baldamus, 1888.
———. "Kopfmessungen kaukasischer Völker." *AA* 18 (1889): 263–281, 298–335; and 19 (1891): 55–84.
Von Luschan, Felix. "Die anthropologische Stellung der Juden." *Correspondenzblatt der deutschen Gesellschaft für Anthropologie, Ethnologie und Urgeschichte* 23 (1892): 94–102.
———. "Offener Brief an Herrn Dr. Elias Auerbach." *ARGB* 4, no. 3 (1907): 362–373.
———. *Völker, Rassen, Sprachen: Anthropologische Betrachtungen.* Berlin: Deutsche Buch-Gemeinschaft, 1927.
———. "Zur physischen Anthropologie der Juden." *ZDSJ* 1, no. 1 (1905): 1–4.
Von Maltzan, Heinrich. *Reise in Arabien.* Braunschweig: Vieweg, 1873.
Wacholder, Ben Zion. "Attitudes towards Proselytizing in the Classical Halakha." *Historia Judaica* 20, part 2 (1958): 77–96.
———. "Cases of Proselytizing in the Tosafist Responsa." *Jewish Quarterly Review* 51 (1960–61): 288–315.
Wachter [Georg Heinrich?]. "Bemerkungen über den Kopf des Juden." *Magazin der Gesellschaft naturforschender Freunde für die neuesten Entdeckungen in der gesamten Naturkunde* (1812): 64–65.
Wagenseil, Franz. "Beiträge zur physischen Anthropologie der spaniolischen Juden und zur jüdischen Rassenfrage." *Zeitschrift für Morphologie und Anthropologie* 23 (1925): 33–150.
Wagner, Richard. *Das Judenthum in der Musik.* Offprint. N.p.: n.p., 1942.
———. *Wagner's Prose Works.* Vol. 3. Translated by William Ashton Ellis. New York: Broude Brothers, 1966.
Waldenburg, Alfred. *Das isocephale blonde Rassenelement unter halligfriesen und jüdischen Taubstummen.* Berlin: Calvary, 1902.
Walk, Joseph. *Kurzbiographien zur Geschichte der Juden 1918–1945.* Munich, New York: K. G. Saur, 1988.
Wassermann, Rudolf. "Ist die Kriminalität der Juden Rassenkriminalität?" *ZDSJ* 7, no. 3 (1911): 36–39.
Wateff, S. "Anthropologische Beobachtungen der Farbe der Augen, der Haare und der Haut bei den Schulkindern von den Türken, Pomaken, Tataren, Armenier, Griechen, und Juden in Bulgarien." *Correspondenzblatt der deutschen Gesellschaft für Anthropologie* 34, nos. 7–8 (1903): 58–60.

Weber, F. "Über die Menstrualverhältnisse der Frauen in St. Petersburg." *St. Petersburger Medicinische Wochenschrift* (1883), no. 41: 329–332; no. 42: 338–40; and no. 43: 345–347.

Weerth, C. *Die Entwicklung der Menschen-Rassen durch Einwirkung der Außenwelt.* Lemgo: Meyersche Hof-Buchhandlung, 1842.

Weinberg, Richard. "Das Hirngewicht der Juden." *ZDSJ* 1, no. 3 (1905): 5–10.

———. "Über einige ungewöhnliche Befunde an Judenhirnen." *Biologisches Centralblatt* 23 (1904): 154–162.

———. "Zur Pathologie der Juden." *ZDSJ* 1, no. 8 (1905): 10–11.

———. "Die transkaukasischen Juden." *ZDSJ* 1, no. 5 (1905): 1–4.

Weindling, Paul. *Health, Race and German Politics between National Unification and Nazism, 1870–1945.* Cambridge: Cambridge University Press, 1989.

Weininger, Otto. *Geschlecht und Charakter: Eine prinzipielle Untersuchung.* Munich: Matthes & Seitz, 1980.

———. *Sex and Character.* London: W. Heineman, 1910.

Weisbach, Augustin. "Körpermessungen verschiedener Menschenrassen." *ZE* 9 (1877): Ergänzungsband.

Weiss, Sheila Faith. *Race Hygiene and National Efficiency: The Eugenics of Wilhelm Schallmayer.* Berkeley and Los Angeles: University of California Press, 1987.

Weissenberg, Dorothea. "Die kaukasischen Bergjuden." *MGJV* 10, no. 3 (1908): 122–127; 11, no. 4 (1908): 160–171.

Weissenberg, Samuel. "Aaroniden und Leviten." *ZDSJ* 4, no. 3 (1908): p. 48.

———. "Alte jüdische Grabdenkmäler aus der Krim." *Ost und West* 13 (1913): 230–234.

———. "Anthropometrische Prinzipien und Methoden." *Globus* 89, no. 21 (1906): 350–351.

———. "Armenier und Juden." *AA* 14 (1915): 383–387.

———. "Die autochthone Bevölkerung Palästinas in anthropologischer Beziehung." *ZDSJ* 5, no. 9 (1909): 129–139.

———. "Beiträge zur Frauenbiologie: Die jüdischen rituellen Sexualvorschriften." In *Abhandlungen aus dem Gebiete der Sexualforschung.* Edited by Max Marcuse, vol. 5, no. 2, 5–29. Berlin: A. Marcus & E. Weber, 1927.

———. "Beiträge zur Volkskunde der Juden." *Globus* 77, no. 8 (1900): 130–131.

———. "Beitrag zur Anthropologie der Juden: Aaroniden und Leviten." *ZE* 39, no. 6 (1907): 961–964.

———. "Ein Beitrag zur Anthropologie der Turkvölker. Baschkiren und Meschtscherjaken." *ZE* 24 (1892): 181–215.

———. "Ein Beitrag zur Lehre von den Lesestörungen auf Grund eines Falles von Dyslexie." Inaugural dissertation. Berlin: L. Schumacher, 1890.

———. "Curriculum Vitae." *Akten der medizinischen Fakultät* 1889/90, IIb/Bd. III, 4a, Nr. 135 b. doc. nos. 463 and 464.

———. "Das Feld- und Kejwermessen." *MGJV* 17, no. 1 (1906): 39–45.

———. "Die Fest- und Fasttage der südrussischen Juden in ethnographischer Beziehung." *Globus* 87, no. 15 (1905): 262–271.

———. "Die Formen des ehelichen Geschlechtsverkehrs." *ARGB* 9, no. 5 (1912): 612–616.

———. "Die jemenitischen Juden." *ZE* 41, nos. 3–4 (1909): 309–327.

———. "Eine jüdische Hochzeit in Südrussland." *MGJV* 15, no. 1 (1905): 59–74.

———. "Jüdische Kunst und jüdisches Kult- und Hausgerät." *Ost und West* 3, no. 3 (1903): 202–206.

———. "Jüdische Museen und jüdisches in Museen." *MGJV* 23, no. 3 (1907): 77–88.

———. "Das jüdische Rassenproblem." *ZDSJ* 1, no. 5 (1905): 4–8.

———. "Der jüdische Typus." *Globus* 97, no. 20 (1910): 309–311; and no. 21 (1910): 328–331.

———. "Jüdischer Volkskalender." *MGJV* 16, no. 1 (1913): 1–2.

———. "Jüdische Sprichwörter." *Globus* 77, no. 21 (1900): 339–341.

———. *Jüdische Stammeskunde* (1928). Mimeograph. New York: Leo Baeck Institute.

———. "Die Karäer der Krim." *Globus* 84 (1903): 139–143.

———. "Die kaukasischen Juden in anthropologischer Beziehung." *AA* 8, no. 4 (1909): 237–245.

———. "Kinderfreud und -leid bei den südrussischen Juden." *Globus* 83, no. 20 (1903): 315–320.

———. "Krankheit und Tod bei den südrussischen Juden." *Globus* 91, no. 23 (1907): 357–363.

———. "Die mesopotamischen Juden in anthropologischer Beziehung." *AA* 10, no. 2–3 (1911): 233–239.

———. "Das neugeborene Kind bei den südrussischen Juden." *Globus* 93, no. 6 (1908): 85–88.

———. "Palästina in Brauch und Glauben der heutigen Juden." *Globus* 92, no. 17 (1907): 261–264.

———. "Peki'in und seine Juden." *Globus* 96, no. 3 (1909): 41–45.

———. "Die persischen Juden in anthropologischer Beziehung." *ZDSJ* 7, no. 1 (1911): 1–6.

———. "Das Purimspiel von Ahasverus und Esther." *MGJV* 13, no. 1 (1904): 1–26.

———. "Rothschild-Legenden." *MGJV* 16, no. 1 (1913): 8–9.

———. "Die Samaritaner." *Ost und West* 13 (1913): 679–690.

———. "Die Spaniolen: Eine anthropometrische Skizze." *MAGW* 39 (1909): 225–236.

———. "Speise und Gebäck bei den südrussischen Juden in ethnologischer Beziehung." *Globus* 89, no. 2 (1906): 25–30.

———. "Südrussische Amulette." *Verhandlungen der Berliner Gesellschaft für Anthropologie, Ethnologie und Urgeschichte* (1897): 367–369.

———. "Die südrussischen Juden: Eine anthropometrische Studie." *AA* 23 (1895): 347–423, 531–579.

———. "Das Sukkothfest in Südrussland." *MGJV* 15, no. 1 (1912): 6–11.

———. "Die syrischen Juden anthropologisch betrachtet." *ZE* 43, no. 1 (1911): 80–90.

———. "Das Wachstum des Menschen nach Alter, Geschlecht und Rasse." *Globus* 94, no. 7 (1908): 101–109.

———. "Weihnachtskerzen—Chanukahkerzen." *MGJV* 15, no. 2 (1912): 71–72.

———. "Die zentralasiatischen Juden in anthropologischer Beziehung." *ZDSJ* 5, no. 7 (1909): 103–106.

———. "Die zentralasiatischen Juden in anthropologischer Beziehung." *MAGW* 43 (1913): 257–269.

———. "Zur Anthropologie der deutschen Juden." *ZE* 44 (1912): 269–274.

———. "Zur Anthropologie der nordafrikanischen Juden." *MAGW* 42 (1912): 85–102.

———. "Zur Anthropologie der persischen Juden." *ZE* 45 (1913): 108–119.

———. "Zur Biotik der südrussischen Juden." *ARGB* 9, no. 2 (1912): 200–206.

———. "Zur Sozialbiologie und Sozialhygiene der Juden." *ARGB* 19, no. 4 (1927): 402–418.

Wertheimer, Jack L. "The 'Ausländerfrage' at Institutions of Higher Learning: A Controversy over Russian Jewish Students in Imperial Germany." *LBIYB* 27 (1982): 187–215.

———. "Between Tsar and Kaiser: The Radicalisation of Russian Jewish University Students in Germany." *LBIYB* 28 (1983): 329–349.

———. "German Policy and Jewish Politics: The Absorption of East European Jews in Germany, 1868–1914." Ph.D. diss., Columbia University, 1978.

———. *Unwelcome Strangers: East European Jews in Imperial Germany.* New York: Oxford University Press, 1987.

Westermarck, Edward. *The History of Human Marriage,* 3 vols. London: Macmillan, 1921.

Williams, Robert G. *Culture in Exile: Russian Emigres in Germany, 1881–1941.* Ithaca: Cornell University Press, 1972.

Wilson, Stephen. *Ideology and Experience: Anti-Semitism in France at the Time of the Dreyfus Affair.* New Jersey: Fairleigh Dickinson University Press, 1982.

Wininger, S. *Grosse jüdische National-Biographie.* 7 vols. Czernowitz: Arta, 1927–1936.

Wiseman, D.J. *Peoples of the Old Testament.* Oxford: Clarendon Press, 1973.

Wistrich, Robert S. *The Jews of Vienna in the Age of Franz Joseph.* Oxford: Oxford University Press, 1989.

———. *Socialism and the Jews: The Dilemmas of Assimilation in Germany and Austria-Hungary.* Rutherford, N.J.: Farleigh Dickinson University Press, 1982.

Wolf, Immanuel. "On the Concept of a Science of Judaism (1822)." *LBIYB* 2 (1957): 194–204.

Yerushalmi, Yosef Hayim. *Assimilation and Racial Anti-Semitism: The Iberian and the German Models*. Leo Baeck Memorial Lecture, no. 26 (1982).

———. *From Spanish Court to Italian Ghetto: Isaac Cardoso: A Study in Seventeenth-Century Marranism and Jewish Apologetics*. Seattle: University of Washington Press, 1981.

Zimmermann, Moshe. "Jewish Nationalism and Zionism in German-Jewish Students' Organizations." *LBIYB* 27 (1982): 129–153.

Zmarlik, Hans-Günther. "Antisemitismus im Deutschen Kaiserreich." In *Die Juden als Minderheit in der Geschichte*. Edited by Bernd Martin, and Ernst Schulin, 249–270. Munich: DTV, 1981.

Zollschan, Ignaz. *Jewish Questions: Three Lectures*. New York: Bloch, 1914.

———. "The Jewish Race Problem." *Jewish Review* 2, no. 11 (1912): 391–408.

———. *Racialism against Civilization*. London: New Europe, 1942.

———. *Das Rassenproblem unter besonderer Berücksichtigung der theoretischen Grundlagen der jüdischen Rassenfrage*. Vienna: Wilhelm Braumüller, 1912.

———. *Revaluation of Jewish Nationalism: A Sociological Study*. London: IGUL Zionist Fraternities of Austrian Universities, 1943.

Zunz, Leopold. *Die gottesdienstlichen Vorträge der Juden historisch entwickelt: Ein Beitrag zur Altertumskunde und biblischen Kritik, zur Literatur- und Religionsgeschichte*. Hildesheim: Georg Olms, 1966.

———. "Grundlinien zu einer künftigen Statistik der Juden." *Zeitschrift für die Wissenschaft des Judenthums* 1 (1822): 523–532.

INDEX

Abortion, 142, 209n46
Acculturation, 125, 132
Adair, James, 37
Alcoholism, 26–27, 159, 178
Allgemeine Zeitung des Judentums, 145
Alsace-Lorraine, 191n22
Alsacian Jews, 111
Alsberg, Moritz, 26
Amorites, 114, 137–38
Amsterdam, 68
Amulets, 108–9, 200n44
Andree, Richard, 21–22
Anthropologie und Zionismus (Sandler), 123
Anthropology, 11; history of, 2–3, 8–9, 13–14, 74; treatment of Jews, 3, 11, 20, 24–25, 26; Jewish scientists and, 8, 9, 29, 64, 97; classification of racial types, 14, 15–16, 22, 24; use of history in, 128–29; Zionism and, 163. *See also* British anthropology; German anthropology; Jewish race science; Race science
Anthropology, cultural, 2, 108–9, 113, 166, 180
Anthropology, physical, 11, 179; history of, 14, 15, 16, 55, 91, 157, 166, 180; Weissenberg's study of, 92, 95, 96, 97, 102–3, 113–14, 121; Zollschan and, 157
Anthropology, social, 2, 180
Anthropometry, 180; surveys by Jacobs, 84, 85, 86–88, 195n76; surveys by Weissenberg, 95, 96, 98–99, 101–2, 104, 115, 117, 118, 178. *See also* Craniometry
Antiquities of Mexico (Kingsborough), 37
Antisemitism: scientific, 5, 6–7, 94, 175–76; in Middle Ages, 5–6, 70–71, 94, 96, 175; race laws, 6, 7; Jewish assimilation and, 6, 7, 22, 52; in Germany, 7, 9, 11, 16, 21–22, 28, 31–32, 52, 58–59, 68, 93,

243

Antisemitism: scientific (*continued*) 94–95, 124, 125, 197*n*2; European, 7, 9, 72, 74, 175–76; in Austria, 7, 61, 154, 197*n*2; Jewish race science and, 7–8, 153, 171–72, 176, 179, 180; in France, 10, 68, 130; Zollschan and, 18, 153, 154–56, 162, 163, 165, 166; in medical profession, 31–32; in literature and arts, 34, 49; in England, 34, 52, 56; and Jewish occupations, 48, 67, 68, 74; Jacobs and, 58, 59, 62, 63, 189*n*9, 189–90*n*10; Weissenberg and, 94, 95, 103, 108, 109, 113, 125; in Russia, 103; Nazi, 104, 165–66, 180; Zionism and, 124, 125, 126–27, 131, 164, 165; and Diaspora Jewry, 127; in Vienna, 154–56; in United States, 180
Arabs: relation to Jews, 86, 105, 115, 139; Zionism and, 140, 173
Archiv für Rassen- und Gesellschaftsbiologie, 19, 127
Arendt, Hannah, 212*n*87
Armenians, 119–20
Aryan racial theory, 88, 157, 159, 160–62, 166
Aryans, 18, 26
Ashkenazic Jews: French emancipation of, 10; distinguished from Sephardim, 24, 82, 88, 104, 114, 117–19, 179; conflated with Sephardim, 56, 157–58
Assimilation: and Iberian antisemitic laws, 6; European antisemitism and, 7; German antisemitism and, 22, 52; Jewish race science and, 29, 176, 179, 180; Knox and, 50, 52, 54; Jacobs and, 63, 69, 84; Weissenberg and, 108, 110, 112, 113; and Jewish art, 112; Zionism and, 124, 125, 135, 136, 158, 164; Auerbach and, 135, 136, 138; Theilhaber and, 149, 153, 211*n*70; Zollschan and, 156, 158, 163, 164
Association for Jewish Statistics, 167, 168–69
Auerbach, Elias, 158, 206–7*n*19; as race scientist, 8, 131, 141; and Jewish racial purity, 127–30, 132, 134, 136, 137, 139–40; and intermarriage, 128, 129–30, 131, 132, 135; commitment to Zionism, 128, 135, 136, 139–41, 173; education and background, 128, 206*n*18, 207*n*20, 207–8*n*27; on economic role of Jews, 130, 131, 132; and Jewish proselytism, 132, 133–34; and Jewish Diaspora, 134, 135–36; and assimilation, 135, 136, 138; study of Palestine Jews, 136–39; and von Luschan's Hittite theory, 137, 138–39, 140–41; and Semitic type, 138–39; *Palästina als Judenland*, 140; "Die Jüdische Rassenfrage," 141
Australia, 60
Austria: antisemitism in, 7, 61, 154, 197*n*2; Jewish marriage and childbirth rates, 75, 194*n*62; military service in, 99–100
"Auto-emancipation," 169

Ba'al Shem Tov. *See* Israel ben Eliezer
Bacharach, Walter Zwi, 165–66
Baratz, H. M., 199*n*37
Baron, Salo W., 72
Beddoe, John, 55–56
Bedouins, 105
Berlin, 151
Bernstein, Ignatz, 111, 200*n*50
Bey, Pruner, 138

Bible, 41, 137, 147
Billroth, Theodor, 30–31, 156
Biology, 13, 175
Birthrates, 75; Jacobs and, 66, 76, 78–80, 81, 194*n*62; Auerbach and, 129; German Jews, 149; Theilhaber and, 150, 151, 152, 153; Ruppin and, 194*n*61
Blacks, 4, 14–15, 44; Jews as, 50–52
Blechmann, Bernhard, 23–24
Blood libel, 71
Blumenbach, Johann Friedrich, 14, 20, 107, 177, 183*n*5
Boas, Franz, 86, 166, 179–80
Bodenheimer, Max, 204–5*n*8
Boulainvilliers, Henri de, 212*n*87
British anthropology, 180; lack of interest in Jews, 11, 33, 34–36, 41, 56–57, 58; treatment of Jews in, 40, 44, 45, 50, 54–55
Broca, Paul, 56
Buchanan, Rev. Claudius, 39–41, 43
Buffon, Georges Louis Leclerc, comte de, 42
Bureau for Jewish Statistics, 167, 172–73, 213*n*105
Buschan, Georg, 185*n*41

Camper, Peter, 14
Cantimpré, Thomas de, 5–6
Capitalism, 130, 150–51
Caucasian Jews, 106–7, 116, 117
Caucasians, 14–15
Central Asian Jews, 116–17, 203*n*72
Centralverein deutscher Staatsbürger jüdischen Glaubens (CV), 10, 31, 127, 146, 206*n*15
Chamberlain, Houston Stewart, 52, 141–42, 155, 189–90*n*10
Chesterton, G. K., 189*n*9
Children: mortality rates, 70–71; physical health, 100–101; of mixed marriages, 129. *See also* Birthrates
"Chinesism," 69–70
Christianity: Enlightenment and, 14; conversion to, 37, 133, 149; and myths of origin, 38; Knox and, 48; relation to Judaism, 63, 116; anti-Jewish beliefs, 70–71, 172; science and, 103
Christian Researches in Asia (Buchanan), 39–40
Christian Social party (Germany), 9
Circumcision, 43, 44
Clermont-Tonnerre, Count Stanislas de, 18
Clifford, James, 2, 3
Clifford, William Kingdon, 189*n*8
Climate, 6, 41–42, 43, 44–45, 82
Colonization, 53–54
"Comparative Distribution of Jewish Ability" (Jacobs), 74–75, 162
Coningsby (Disraeli), 48–49
"Consanguineous Marriages" (Jacobs), 65
Contraception, 194*n*61, 203–4*n*82
Craniometry: cephalic index, 22; and Jewish racial dualism, 22–23, 24; Jacobs and, 74, 84, 85–88; and intellectual ability, 74, 87–88; Auerbach and, 139. *See also* Anthropometry
Criminality, 172
Curtin, Philip D., 187*n*31

Daniel Deronda (Eliot), 60
Darwin, Charles R., 61, 66, 191*n*24
Darwin, Sir George H., 65
Darwinism, 61
Decolonization, 3
De generis humani varietate nativa (Blumenbach), 20
Diaspora Jews: Zionism and, 12, 121, 125, 134, 143, 144, 173;

Diaspora Jews (*continued*)
and Jewish racial varieties, 24, 39, 87; Jacobs and, 87; Weissenberg and, 92, 121; antisemitism and, 127; Auerbach and, 134, 135–36
Dietary laws, 132, 201*n*55
Disease and illness: Jewish susceptibility to, 26, 27, 81, 83
Disraeli, Benjamin, 48–49, 50, 187–88*n*38
Dreyfus affair, 10
Drumont, Eduard, 10

East European Jews: immigration to Germany, 9, 21–22, 93–94, 149, 150, 152; German antisemitism and, 9, 22, 30, 93; anthropometric classifications of, 23–24, 86, 99, 117; immigration to England, 35, 36, 56, 62–63, 88; in England, Jacobs and, 69, 77–78, 88; Weissenberg's study of, 102, 105, 107–9, 110, 114, 117, 120, 203–4*n*82; folk culture of, 108, 109, 111, 113; Zionism and, 152, 204–5*n*8; immigration to Vienna, 156; Jewish race science and, 178, 179. *See also* Russian Jews
Education, 30, 93–94, 100–101, 152
Edward I (king of England), 34
Edwards, William Frederick, 184*n*23
Eliot, George, 60–61
Emancipation: and Jewish identity, 4, 153; race science and, 5, 29, 176; in France, 10; in Germany, 10, 17, 21, 52; in England, 34, 52–53
Endelman, Todd M., 187–88*n*38
England: expulsion of Jews, 34; antisemitism in, 34, 52, 56; eastern Jewish immigration to, 35, 36, 56, 62–63, 88; myth of Hebrew origins in, 38, 39; revolutions of 1848 and, 45, 47;
National Efficiency movement, 56, 75; Jacobs and, 61, 62, 88; census of, 65; commerce in, 72. *See also* British anthropology
English Jews: British anthropology and, 11, 33, 35, 45, 56–57; mental health of, 27; British acceptance of, 33, 34, 35, 36, 56–57, 58, 162, 177, 186*n*9; British animosity toward, 33–34; emancipation of, 34, 52–53; environmental influence and, 43, 195*n*76; assimilation of, 52; population, 54; marriage and childbirth patterns, 65–66, 78; poverty among, 69; infant mortality rates, 70; occupations of, 72, 73; Jacobs and, 78, 88, 195*n*76
Enlightenment: and environmental influence, 6, 160; and race science, 8–9, 13, 14, 160, 175; and Jews and Judaism, 38, 59
Environmental influence: Grégoire and, 6; Jewish race science and, 9, 160, 173, 176–77; Zionism and, 12, 173; Vogt and, 23; Prichard and, 43; Jacobs and, 59, 64–65, 85, 86, 159–60; Weissenberg and, 95, 97, 98, 100, 101; Zollschan and, 157, 159–61
Equality of the European Races, 164–65
Essay on the Causes of the Variety of Complexion and Figure in the Human Species (Smith), 42
Essay on the Physical, Moral, and Political Regeneration of the Jews (Grégoire), 6
Ethnography, 2, 3
Ethnological Society of London, 45–46
Eugenics, 19, 120, 148, 179, 196*n*89

INDEX

Europe: anthropology in, 3, 24; antisemitism in, 7, 9, 72, 74, 175–76; racialism in, 13, 15–16, 24, 45, 47–48, 55, 126, 176; national myths of origin, 37–38; revolutions of 1848, 45, 47, 51; nationalism in, 110, 124, 154; Zionist attitudes toward, 126, 140; Jewish contributions to, 126, 162
European Jews, 3–4, 55, 107; emancipation of, 5, 176; legislative restrictions against, 6, 7; assimilation of, 7, 54; physical appearance of, 21, 50–51; Jacobs's demographic study of, 67, 75, 78, 80–81, 90; urbanization of, 72. *See also* East European Jews; West European Jews

Facial angle, 14–15
Fischer, Eugen, 19
Fishberg, Maurice, 133, 192*n*29, 195*n*77
Flinders Petrie, W. M., 138
Foundations of the Nineteenth Century (Chamberlain), 141–42, 155, 189–90*n*10
France, 212*n*87; Jewish emancipation, 10; antisemitism in, 10, 68, 130; race science in, 10–11; marriage and birthrates in, 75, 78
Franco-Prussian War, 10
Franzos, Karl Emil, 109
French Jews, 10–11, 54
French revolution, 149
Freud, Sigmund, 206–7*n*19

Galician Jews, 195*n*76
Galton, Sir Francis, 62, 74, 81, 88, 196*n*89
Garcia, Gregorio, 37

General History of Civilization (Klemm), 15
German anthropology, 14, 94–95; treatment of Jews in, 11, 16–18, 20, 23, 24–25, 26, 28, 35; antisemitism in, 18–19, 94
German Jews: self-definitions, 4, 109, 110–11; and eastern Jewish immigrants, 9, 22, 108; emancipation of, 10, 17, 21, 52; self-defense organizations, 10, 126–27; considered outside German society, 16–17, 21, 25, 28, 30–31, 35, 172; population, 16, 149–50; assimilation of, 22, 52, 124, 125, 132, 136, 138, 188*n*44; religious practices, 27–28, 178; physicians, 30, 31; rejection of nationalism, 109, 110; Yiddish-speaking, 111; Zionist theories of demise of, 124–25, 132, 136, 141–42, 144–47, 149–50, 152, 153, 170, 174; intermarriage, 129, 131, 138; in capitalist industry, 150–51; and Jewish scholarship, 167
Germany: antisemitism in, 7, 9, 11, 16, 21–22, 28, 31–32, 52, 58–59, 68, 93, 94–95, 124, 125, 197*n*2; eastern Jewish immigration to, 9, 21–22, 93–94, 149, 150, 152; unification of *1871*, 16, 28; "Jewish racial question" in, 17–18, 32, 177; racial hygiene movement, 18, 19; racial surveys of, 25; nationalism in, 28–29, 94–95, 136; higher education, 30, 93–94; exclusionary laws, 130
Geschlecht und Charakter (Weininger), 96
Gliddon, George R., 41
Gobineau, Joseph Arthur, comte de, 45, 46–47, 187*n*30

247

INDEX

Golden, Peter B., 199*n*37
Goldmann, Felix, 206*n*15
Gordin, Jacob, 197*n*7
Graetz, Heinrich, 52–53, 148
Grattenauer, Karl Wilhelm Friedrich, 175
Grégoire, Abbé Henri Baptiste, 6, 7, 175, 182*n*11
Grusians, 116

Halakha. *See* Jewish law
Hebrew language, 110, 140, 201*n*56
Hereditary Genius (Galton), 74
Heredity, 9, 74, 82, 161, 173
Herzen, Alexander, 99
Herzl, Theodor, 126, 156, 205*n*11, 210*n*63
Hirschfeld, Magnus, 143
"Historical Notes on the Blood Accusations" (Jacobs), 71
History, 128–29
History of the American Indians (Adair), 37
History of the European Mind (Jacobs), 89–90
History of the Jewish Homily (Zunz), 167
Hitler, Adolf, 154, 165, 180
Hittites, 119; relation to Jews, 26, 86, 114, 115, 137
Hofacker, Johann Daniel, 80
Hope of Israel (Menasseh ben Israel), 36–37
Hoyzfraynd, 111
Hungary, 75
Huxley, Henry Minor, 202*n*64
Huxley, Thomas Henry, 189*n*8

Iewes in America (Thorowgood), 37
Ikow, Constantine, 24
Im deutschen Reich, 31, 146
India, 40, 41
Infant mortality, 70

Innitzer, Theodor, Cardinal, 165
Insanity, 27
Intellectual ability: Jacobs and, 73–75, 88, 159–60, 162; Zollschan and, 160, 161–62
Intermarriage: fertility of offspring, 8, 79–80; and Jewish physical appearance, 24, 42, 56, 82, 85–86; childbirth rates in, 79; Jacobs and, 79–80, 81; proscriptions of, 87, 132, 171; Weissenberg and, 119–20; Auerbach and, 128, 129–30, 131, 132, 135; Theilhaber and, 129, 150, 153; children of, 129, 207–8*n*27; Zionism and, 145, 149, 170; Zollschan and, 163, 164
International Hygiene exhibition (1911), 148
Israel ben Eliezer (Ba'al Shem Tov), 109
Israelit, Der, 145
Israelitische Wochenschrift, 145

Jacobs, Joseph: as race scientist, 8, 9, 11, 57, 58–60, 90, 91, 92, 143; responses to antisemitism, 58, 59, 62, 63, 189*n*9, 189–90*n*10; and environmental influence, 59, 64–65, 85, 86, 159–60; and Jewish racial purity, 59, 82, 86, 97–98, 178; education and background, 60–62; and Jewish history, 61–62, 63–64, 90; and Jewish contributions to civilization, 62, 75, 90; working relationship with Galton, 62, 196*nn*89, 90; and Russian Jews, 62–63, 108, 125, 190*n*13, 195*n*76; and assimilation, 63, 69, 84; statistical surveys of Jews, 64, 65, 80–81, 172, 190–91*n*19, 213*n*105; study of consanguineous marriages, 65–67;

248

statistics on marriage and birthrates, 66, 75–77, 78–80, 81, 178, 194*n*62, 195*n*73; study of Jews' occupations, 67–68, 69, 71–73, 75, 193*n*43; statistics on wealth and poverty, 68–69, 79, 81, 84; and eastern Jews in England, 69, 77–78, 88; on "Chinesism," 69–70; and Judaism, 70, 73, 191–92*n*28; on blood libel, 70–71; study of intellectual ability, 73–75, 88, 159–60, 162; on longevity and diseases, 82–84; anthropometric and craniometric surveys, 84, 85, 86–88, 159–60, 195*n*76; as *Jewish Encyclopaedia* editor, 88–89; on physical appearance of Jews, 97–98, 138; and Ruppin's *Juden der Gegenwart*, 213*n*109. Works, 189*n*2, 190*nn*12, 16; *Jewish Statistics*, 63, 190*n*18; "Studies in Jewish Statistics," 64–65; "Consanguineous Marriages," 65; "The Social Condition of London Jewry," 67; "Historical Notes on the Blood Accusations," 71; "Little St. Hugh of Lincoln," 71; "Occupations," 71; "Occupations of London Jews," 71; "Professions," 71, 73–74; "Comparative Distribution of Jewish Ability," 74–75, 162; *Jewish Contributions to Civilization*, 75; *Men of Distinction*, 75; "Vital Statistics," 75; "On the Racial Characteristics of Modern Jews," 82–83, 90; *Jewish Encyclopaedia* entries, 88–89; *The Jewish Race*, 89; *The History of the European Mind*, 89–90; *Jews of Distinction*, 196*n*90

Jellinek, Adolf, 96
Jewish art, 111–12
Jewish Chronicle, 64, 65–66
Jewish Contributions to Civilization (Jacobs), 75
Jewish Encyclopaedia, 88–89
Jewish folklore, 110–11
Jewish history, 177; Jacobs and, 61–62, 63–64, 90; Zionism and, 121, 134–35, 174; Auerbach and, 134, 135–36
Jewish identity: Jewish race science and, 4, 7, 29, 31, 177–78; emancipation and, 4, 153; Zionism and, 12, 125–26, 135
Jewish law: and sexuality, 132, 147–48; dietary laws, 132, 201*n*55
Jewish nationalism, 110, 112, 126, 136. *See also* Zionism
Jewish Race (Jacobs), 89
Jewish race science, 3, 4–5, 29–30, 31–32, 180; and Jewish identity, 4, 7, 29, 31, 177–78; and antisemitism, 7–8, 153, 171–72, 176, 179, 180; appropriation of anthropological methodology, 7–10, 12, 29, 119, 178–79; and Enlightenment values, 8–9, 160; and environmental influence, 9, 160, 173, 176–77; Zionism and, 11–12, 76, 78, 124–27, 173; and "Jewish racial question," 29, 32, 173, 176–77; and assimilation, 29, 176, 179, 180; Jacobs and, 57, 58–60, 63, 76, 89, 90, 178; Weissenberg and, 91–92, 97, 118, 121, 178; Zollschan and, 153, 160, 178; Zunz and, 166–67
Jewish racial purity: German anthropology and, 20, 24; Jacobs and, 59, 82, 86, 97–98, 178; Weissenberg and, 97, 98, 104–5, 114–15, 116; Zollschan and, 97,

Jewish racial purity (*continued*)
157–60, 163; Auerbach and, 127–30, 132, 134, 136, 137, 139–40
"Jewish racial question," 3–4, 19–20; in France, 10–11; in Germany, 17–18, 32, 177; Jewish race scientists and, 29, 32, 173, 176–77; Jacobs and, 91; Zollschan and, 157
Jewish Statistics (Jacobs), 63
Jewish *Urtypus*, 92–93, 102, 105, 112, 113–14, 117, 118, 119, 178, 201–2n61
Jews. See names of nationalities of Jews
Jews and Modern Capitalism (Sombart), 130
Jews of Distinction (Jacobs), 196n90
"Jews of Southern Russia" (Weissenberg), 95, 96–97
Judaism, 27–28; influence of, 46; relation to Christianity, 63, 116; Jacobs and, 70, 73, 191–92n28; and occupations of Jews, 73, 193n47; conversion of Khazars to, 106, 133, 199n37; Weissenberg and, 120–21; relation to Islam, 132; conversion to, 132–34; and reproduction, 148–49; Zionism and, 178
Juden der Gegenwart, Die (Ruppin), 168, 194n61
Judenthum in der Musik (Wagner), 49
"Jüdische Rassenfrage, Die" (Auerbach), 141

Kautsky, Karl, 124, 174
Khazars, conversion to Judaism, 106, 133, 199n37
Kingsborough, Viscount Edward King, 37
Klausner, Joseph, 134–35

Klemm, Gustav, 15
Knox, Robert, 46; *The Races of Men*, 45, 47, 48, 51, 54–55; racial view of history, 47–48, 53–54, 56, 187n31; and Disraeli, 48, 50, 52; and assimilation, 50, 52, 54; on physical appearance of Jews, 50–51
Kollmann, Julius, 24
Kraepelin, Emil, 27, 182n13
Krafft-Ebing, Richard von, 27, 28, 185n44

Lagneau, Gustave, 24, 80
Langbehn, Julius, 188n44
Language and race, 157
Lassalle, Ferdinand, 52
Latham, Robert Gordon, 45–47
Laubhütte, Die, 145
Lawrence, Sir William, 44–45
Lazarus, Moritz, 71
Lectures on Man (Vogt), 23
Lectures on Physiology, Zoology, and the Natural History of Man (Lawrence), 44
Lenz, Fritz, 19
Leopold, Louis, 169
Lewy, Joseph, 31–32
"Little St. Hugh of Lincoln" (Jacobs), 71
Loewe, Heinrich, 207–8n27, 128
Lombroso, Cesare, 86
London, England, 67; Jewish settlement in, 33, 35, 36; Jewish poverty and prosperity in, 68–69; Jewish occupations in, 69, 72, 73; Jewish health in, 70, 71
Longevity, 81, 82–83
Lonsdale, Henry, 187n31
Lueger, Karl, 155, 156, 157

Marriage: consanguineous, 65–67, 191nn22, 24; rates of, 75–76; age

at, 76, 77, 151; childbirth rates in, 79–80, 178; proselytism requirement, 87; and physical health, 101; and racial purity, 134, 138; monogamy requirement, 191n21. *See also* Intermarriage
Marx, Karl, 52
Masaryk, Thomas, 164
Maskilim, 101
Maurer, Friedrich, 23
Medicine, 11, 12, 15; folk medicine, 109. *See also* Anthropology; Race science
Meiners, Christoph, 14
Menasseh ben Israel, Rabbi, 36–37
Mendelssohn, Moses, 17, 63
Men of Distinction (Jacobs), 75
Menstruation, 5–6, 76–77, 178
Michaelis, Curt, 170–71
Middle Ages: antisemitic beliefs in, 5–6, 70–71, 94, 96, 175; persecution of Jews in, 94, 103, 105, 133; intermarriage in, 118, 129–30
Military service, 99–100, 195n77, 198nn21, 22
Millenarianism, 36–37
Miscegenation. *See* Intermarriage
Montezinos, Antonio de, 36–37
Mosse, George L., 12, 25, 102, 187n30
Mountain Jews, 116, 117
Multiculturalism, 181–82n3
Music, 49, 195n73
Muskeljudentum (Nordau), 150

Napoleon I (emperor of France), 65
Nationalism: influence on anthropology, 22; German, 28–29, 94–95, 136; Jewish, 110, 112, 126, 136; European, 110, 124, 154
Natural History of the Varieties of Man (Latham), 45

Nazism, 2, 104, 155, 165–66, 179
Neubauer, Adolf, 82
Nordau, Max, 150, 168–69
North African Jews, 119
Nossig, Alfred, 167, 168, 170, 171, 172
Nothnagel, Hermann, 31
Nott, Josiah C., 41

Occupations: Knox and, 48; Jacobs's study of, 67–68, 69, 71–73, 75, 193n43; Judaism-oriented, 73, 193n47; Weissenberg and, 101; Bureau for Jewish Statistics and, 172
"Occupations" (Jacobs), 71
"Occupations of London Jews" (Jacobs), 71
"On the Racial Characteristics of Modern Jews" (Jacobs), 82, 90
Origen de los Indios (Garcia), 37
Orleans, Council of (538), 133
Ost und West, 146
"Outline for a Future Statistics of the Jews" (Zunz), 166

Palästina als Judenland (Auerbach), 140
Palestine, 144; Weissenberg's study of Jews of, 114, 115, 117, 202n65; Auerbach's study of Jews of, 136–39; Zionism and Arabs of, 140, 173
Paxton, John, 47
Persian Jews, 116, 117
Physicians, 30, 31
Physiognomy: of Jews, 6, 20–21, 50, 86, 103–4; facial angle, 14–15. *See also* Craniometry
Piacenza, Franco da, 6
Pinsker, Leo, 169
Pius XI, 165
Ploetz, Alfred, 18

Pobedonostsev, Konstantin, 103
Pogroms, 21–22, 61, 63
Poland, 130
Polish Jews, 195n76
Politisch-Anthropologische Revue, 18–19
Polygamy, 191n21
Polygenism, 40–41, 79
Population genetics, 179
Poverty: Jacobs on, 68, 69, 79, 81, 84; Weissenberg on, 100; Theilhaber on, 152; Zollschan on, 161
"Prepotency," 97–98
Prichard, James Cowles, 41–44, 45–46, 56, 138
Problem of Racial Crossing among Humans (Fischer), 19
"Professions" (Jacobs), 71, 73–74
Proselytes of the Gate, 133
Proselytism, 24, 87, 132–33, 207–8n27
Prussia: discrimination against Jews in, 34, 61; Jewish poverty in, 68; Jewish demographics, 75, 149, 150
Psychiatry, 7, 27–28, 134
Purim, 113, 201n56

Race science, 2, 41, 131, 134; treatment of Jews in, 5, 11, 16, 20; antisemitism in, 5, 57, 176; Jewish scientists' appropriation of, 7–10, 12, 29, 119, 178–79; use of anthropometry and craniometry, 15, 22, 74; physicians as practitioners of, 30; and intellectual capacities, 74; use of eugenics, 120; discrediting of, 141, 179–80. *See also* Anthropology; Jewish race science
Races of Men (Knox), 45, 47, 48, 51, 54–55

"Racial Geography of Europe" (Ripley), 87
Racial pride (*Rassenstolz*), 170–71
Racism, 28, 179, 187n31
Rasse und Judentum (Kautsky), 124
Rassenproblem unter besonderer Berücksichtigung der theoretischen Grundlagen der jüdischen Rassenfrage (Zollschan), 155, 156–57, 160, 166
Rathenau, Walter, 150, 210n63
Reasons for Naturalizing the Jews in Great Britain and Ireland (Toland), 38
Renan, Ernest, 85, 192n35
Researches into the Physical History of Man (Prichard), 41–42
Retzius, Andreas, 22
Ripley, William Z., 87
Roth, Cecil, 199n37
Rothschild, Baron Lionel de, 52
Rudolphi, Karl Asmund, 21
Ruppin, Arthur, 79, 134, 173; and publication of *Jüdische Statistik*, 167–68, 213n108; as *Zeitschrift für Demographie* editor, 168, 170, 214n114; *Die Juden der Gegenwart*, 168, 194n61, 213n109
Russia: pogroms in, 21–22, 61, 63; military service in, 99–100, 198n22; antisemitism in, 103; Jewish settlement in, 106–7; contraceptive use in, 203–4n82
Russian Jews, 24; immigration to Germany, 9, 21–22, 30, 93; and German antisemitism, 22; immigration to England, 36, 62–63, 108; Jacobs and, 62–63, 108, 125, 190n13, 195n76; marriage rates, 75; in German universities, 93–94; Weissenberg's study of, 95, 96–97, 98–99, 100, 101–2, 104–

9, 113, 116–17, 119; Zionism and, 204–5n8. *See also* East European Jews

Sadler, Michael Thomas, 80
Samaritans, 115, 202n64
Sandler, Aron, 123–24, 157, 174, 210–11n68
Scientific racism. *See* Race science
Segall, Jacob, 146, 214n114
Semites, 18–19, 22, 24, 84, 86, 105, 114, 138–39
Sephardic Jews: French emancipation of, 10; distinguished from Ashkenazim, 24, 82, 88, 104, 114, 117–19, 179; conflated with Ashkenazim, 56, 157–58
Sexuality: beliefs about Jews, 6, 7, 27; Jewish law and, 132, 147–48
Smith, Samuel Stanhope, 42, 43
"Social Condition of London Jewry" (Jacobs), 67
Social Statics (Spencer), 47
Sombart, Werner, 130–31, 150, 207n20
Spanish Jews, 131, 132
Spencer, Herbert, 47
Spinoza, Baruch, 60
Sterile Berlin, Das (Theilhaber), 152
Stieda, Ludwig, 24
Stöcker, Adolf, 9
Stöcker, Helene, 142
"Studies in Jewish Statistics" (Jacobs), 64–65
Syrian Jews, 118

Tänzer, Rabbi A., 145–46
Theilhaber, Felix A., 8, 211n69; on birthrates, 79, 150, 151, 152, 153; on intermarriage, 129, 150, 153; *Der Untergang der deutschen Juden*, 141–42, 143, 144–46; education and background, 142, 143; commitment to Zionism, 142, 153; work for reproductive freedoms, 142–43, 209n46; medical practice, 143; theory of German Jewry's demise, 144, 147, 149–53; criticisms directed against, 144–47, 209n49; and Jewish sexual ethic, 147–48; and assimilation, 149, 153, 211n70; *Das Sterile Berlin*, 152
Thorowgood, Thomas, 37
Toland, John, 38–39

United States, 4; immigration to, 86, 149; race science in, 179–80
University of Vienna, 155–56
Untergang der deutschen Juden, Der (Theilhaber), 142, 143, 144–46

Vienna, Austria, 154–56, 211n83
Virchow, Rudolf, 25–26, 31, 137–38, 177, 185n38
"Vital Statistics" (Jacobs), 75
Vogt, Carl, 22–23
Voltaire (François-Marie Arouet), 183n2
Von Luschan, Felix; Hittite theory, 26, 86, 137, 203n80; Weissenberg and, 114–15, 116, 117, 119; Auerbach and, 137, 138–39, 140–41

Wagener, Hermann, 52
Wagner, Cosima, 155
Wagner, Richard, 49
Weerth, C., 15
Weininger, Otto, 96
Weisbach, Augustin, 23
Weissenberg, Samuel A., 197n3, 204n83; as race scientist, 8, 11, 90, 91–93, 118, 120, 121, 143; education and background, 92,

Weissenberg, Samuel A. (*continued*) 93, 94; study of physical anthropology, 92, 95, 96, 97, 102–3, 113–14, 121; and Zionism, 92, 109, 110, 121–22; commitment to Diaspora Jewry, 92, 121; search for Jewish *Urtypus,* 92–93, 102, 105, 112, 113–14, 117, 118, 119, 178, 201–2n61; responses to antisemitism, 94, 95, 103, 108, 109, 113, 125; anthropometric surveys, 95, 96, 98–99, 101–2, 104, 115, 117, 118, 178; "The Jews of Southern Russia," 95, 96–97; study of Russian Jewry, 95, 96–97, 98–99, 100, 101–2, 104–9, 113, 116–17, 119; and environmental influence, 95, 97, 98, 100, 101; and Jewish racial purity, 97, 98, 104–5, 114–15, 116; and Jewish physical strength, 100–101; descriptions of Jewish physical appearance, 102, 103–4; comparison of Ashkenazim and Sephardim, 104, 114, 117–19; on Khazar conversion to Judaism, 106, 133; and assimilation, 108, 110, 112, 113; study of cultural anthropology, 108–9, 110–13, 200n44; and Jewish nationalism, 109–10, 112; on Hebrew language, 110, 201n56; study of Palestine Jews, 114, 115, 117, 202n65; and von Luschan's theories, 114–15, 116, 117, 119; study of Yemenite Jews, 115–16; study of Central Asian Jews, 116–17, 203nn72, 75; study of Syrian Jews, 118; study of North African Jews, 119; study of Armenians, 119–20; criticism of women's movement, 120, 204n84; and Judaism, 120–21; on Jewish dietary laws, 201n55; on contraceptive use, 203–4n82
Welt, Die, 146
Weltsch, Robert, 126
West European Jews, 68, 90, 113, 145, 179, 203–4n82
Wissenschaft des Judentums, 127, 166, 167
Wistrich, Robert S., 154
Woltmann, Ludwig, 18–19
Women: antisemitic beliefs about, 6, 7; race science and, 29, 76–77, 95–96, 143; Weissenberg and, 120, 204n84

Yemenite Jews, 115–16
Yerushalmi, Yosef Hayim, 6
Yiddish, 111

Zangwill, Israel, 189n9, 191–92n28
Zeitschrift für Demographie und Statistik der Juden, 167, 168, 169–70
Zeitschrift für Sexualhygiene, 142–43
Zionism, 205n11; uses of race theory, 9, 11–12, 76, 78, 123, 124, 126, 174; critique of Diaspora Jewry, 12, 121, 125, 127, 143, 173; and Jewish identity, 12, 125–26, 135; and environmental influence, 12, 173; Weissenberg and, 92, 109, 110, 121–22; and Jewish nationalism, 110, 124, 126, 136, 154; and Jewish history, 121, 134–35, 174; responses to antisemitism, 124, 125, 126–27, 131, 164, 165; and assimilation, 124, 125, 135, 136, 158, 164; theories of Jewry's demise, 124–25, 132, 136, 141–42, 144–47, 149–50, 152, 153, 163–64, 170, 174; attitudes toward Europe, 126, 140;

Auerbach and, 128, 135, 136, 139–41, 173; and Arabs of Palestine, 140, 173; Theilhaber and, 142, 153; and east European Jews, 152, 204–5n8; Zollschan and, 154, 156, 163–64, 165; Bureau for Jewish Statistics and, 168; and traditional Judaism, 178

Zionistische Vereinigung fr Deutschland (ZVfD), 126–27

Zollschan, Ignaz: as race scientist, 8, 153, 166, 178; and Jewish contributions to civilization, 12, 156, 162, 163, 178; responses to antisemitism, 18, 153, 154–56, 162, 163, 165, 166; and Jewish racial purity, 97, 157–60, 163; theory of Jewry's demise, 144, 162–64; education and background, 154, 155–56, 211n73; commitment to Zionism, 154, 156, 163–64, 165; opposition to Nazism, 154, 164, 165–66; *Das Rassenproblem*, 155, 156–57, 160, 166; and assimilation, 156, 158, 163, 164; and environmental influence, 157, 159–61; and intellectual ability, 160, 161–62; *Zollschan-Aktion* plan, 164–65, 212n102

Zunz, Leopold, 166–67, 172

Zur Volkskunde der Juden (Andree), 21

Zweig, Arnold, 206–7n19